Technical Communication Fundamentals

William Sanborn Pfeiffer
Warren Wilson College

Kaye E. Adkins
Missouri Western State University

Prentice Hall

Boston Columbus Indianapolis New York San Francisco Upper Saddle River
Amsterdam Cape Town Dubai London Madrid Milan Munich Paris Montreal Toronto
Delhi Mexico City Sao Paulo Sydney Hong Kong Seoul Singapore Taipei Tokyo

Editorial Director: Vernon Anthony
Senior Acquisitions Editor: Gary Bauer
Editorial Assistant: Tanika Henderson
Director of Marketing: David Gesell
Senior Marketing Manager: Leigh Ann Sims
Senior Marketing Assistant: Les Roberts
Project Manager: Holly Shufeldt
Cover Art Director: Jayne Conte
Cover Designer: Suzanne Behnke
Cover Art: iStockPhoto
Full-Service Project Management: Sudip Sinha/Aptara®, inc.
Composition: Aptara®, inc.
Printer/Binder: R.R. Donnelley & Sons
Cover Printer: Lehigh-Phoenix Color

Credits and acknowledgments borrowed from other sources and reproduced, with permission on page 289.

Library of Congress Cataloging-in-Publication Data
Pfeiffer, William S.
   Technical communication fundamentals / William Sanborn Pfeiffer, Kaye E. Adkins.
     p. cm.
   Includes index.
   ISBN-13: 978-0-13-237457-6
   ISBN-10: 0-13-237457-9
  1. English language—Technical English.   2. Communication of technical information.
3. English language—Rhetoric.   4. Technical writing.   I. Adkins, Kaye E.   II. Title.
   PE1475.P468 2011
   808'.0666—dc22
                                                         2010043759

10 9 8 7 6 5 4 3 2 1

Prentice Hall
is an imprint of

ISBN 10:   0-13-237457-9
ISBN 13: 978-0-13-237457-6

Dedication

Deepest thanks go to my family—Evelyn, Zachary, and Katie—for their love and support throughout this and every writing project I take on.

—Sandy

To my family—Perry, Ian, and Evan—for their support and patience during this project.

—Kaye

Preface

>>> Technical Communication Fundamentals *Pfeiffer and Adkins*

From the Authors:

It is a good thing, perhaps, to write for the amusement of the public, but it is a far higher and nobler thing to write for their instruction, their profit, their actual and tangible benefit.

Mark Twain, 1863

When Mark Twain penned this first sentence to "How to Cure a Cold," he could not have envisioned a time when writing for "tangible benefit" would be so much a part of our lives. Because that time has come, this book has been written to help you plan, write, and edit all types of on-the-job writing.

This textbook presents the fundamental concepts, techniques, and genres commonly used in the workplace—supported by useful examples and exercises. We have included the word "fundamentals" in the title deliberately. This text offers a solid starting point for communicating effectively in a future of rapidly changing technology and workplace settings. In contrast, our comprehensive longer text, Technical Communication: A Practical Approach 7e, provides more detailed information about producing technical communication.

Though this book stands on its own as a resource, we have taken advantage of the Web to provide you with many more examples, tutorials, and helpful resources in MyTechcommlab. In the margins of the text pages, you will see icons with notes indicating when something useful is available online in MyTechcommlab. On this ancillary Web site, you will find a variety of additional material, including over 90 model assignments, grammar and editing assistance, research guidelines, writing tutorials, activities and case studies, and additional reference resources in the field of technical communication.

As we wrote in *Technical Communication: A Practical Approach*, we believe that clear, concise, and honest writing can be a powerful tool throughout your working life. If used wisely, it will help you meet challenges you face in landing a job and advancing in your career. We hope this little book moves you toward that goal.

Sandy Pfeiffer,
President, Warren Wilson College

Kaye Adkins,
Associate Professor, Missouri Western State University

>>> **Core Features of** *Technical Communication Fundamentals*

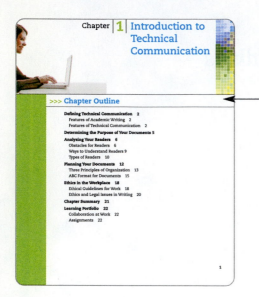

Focus on Process and Product

This book introduces students to writing in the first chapter, and asks them to write throughout the course. The text immerses them in the process of technical writing while teaching practical formats for getting the job done.

A Simple ABC Pattern for All Documents

The "ABC format"—Abstract, Body, and Conclusion—will guide students' work in this course and throughout their careers. This underlying three-part structure provides a convenient handle for designing almost every technical document.

Numbered Guidelines

Many sets of short, numbered guidelines make this book easy to use to complete class projects. Each set of guidelines will take students through the process of finishing assignments, such as writing a proposal, doing research on the Internet, constructing a bar chart, and preparing an oral presentation.

Annotated Models

The text contains models grouped at the end of chapters on color-edged pages for easy reference. Annotations in the margins are highlighted in color and show exactly how the sample documents illustrate the guidelines set forth in the chapters.

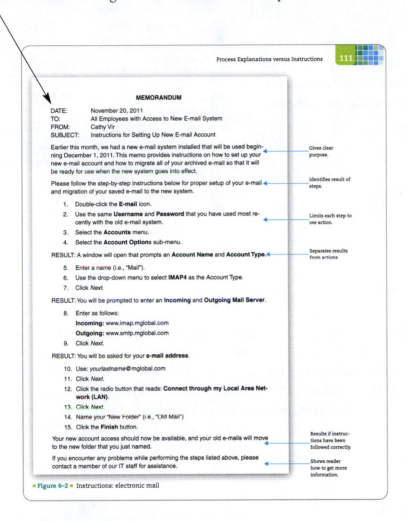

MEMORANDUM

DATE: November 20, 2011
TO: All Employees with Access to New E-mail System
FROM: Cathy Vir
SUBJECT: Instructions for Setting Up New E-mail Account

Earlier this month, we had a new e-mail system installed that will be used beginning December 1, 2011. This memo provides instructions on how to set up your new e-mail account and how to migrate all of your archived e-mail so that it will be ready for use when the new system goes into effect.   *Gives clear purpose.*

Please follow the step-by-step instructions below for proper setup of your e-mail and migration of your saved e-mail to the new system.   *Identifies result of steps.*

1. Double-click the **E-mail** icon.
2. Use the same **Username** and **Password** that you have used most recently with the old e-mail system.   *Limits each step to one action.*
3. Select the **Accounts** menu.
4. Select the **Account Options** sub-menu.

RESULT: A window will open that prompts an **Account Name** and **Account Type.**   *Separates results from actions.*

5. Enter a name (i.e., "Mail").
6. Use the drop-down menu to select **IMAP4** as the Account Type.
7. Click *Next*.

RESULT: You will be prompted to enter an **Incoming** and **Outgoing Mail Server.**

8. Enter as follows:
   Incoming: www.imap.mglobal.com
   Outgoing: www.smtp.mglobal.com
9. Click *Next*.

RESULT: You will be asked for your **e-mail address.**

10. Use: *yourlastname*@mglobal.com
11. Click *Next*.
12. Click the radio button that reads: **Connect through my Local Area Network (LAN)**.
13. Click *Next*.
14. Name your "New Folder" (i.e., "Old Mail")
15. Click the **Finish** button.

Your new account access should now be available, and your old e-mails will move to the new folder that you just named.   *Results if instructions have been followed correctly.*

If you encounter any problems while performing the steps listed above, please contact a member of our IT staff for assistance.   *Shows reader how to get more information.*

■ Figure 6–2 ■ Instructions: electronic mail

mytechcommlab Additional Sample Documents in MyTechCommLab

In the margins of this book, you will find the MTCL icon whenever there are additional sample documents or other useful tools available for use online.

>>> Building Your Technical Communication Skills

Collaboration at Work

Every chapter also includes a "Collaboration at Work" exercise. These exercises engage the student's interest in chapter content by getting teams to complete a simple project.

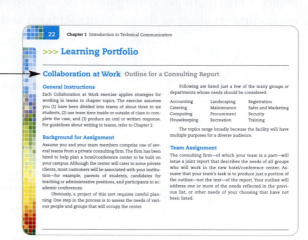

22   Chapter 1 Introduction to Technical Communication

>>> Learning Portfolio

Collaboration at Work Outline for a Consulting Report

General Instructions

Each Collaboration at Work exercise applies strategies for working in teams to chapter topics. The exercise assumes you (1) have been divided into teams of about three to six students, (2) use team time inside or outside of class to complete the case, and (3) produce an oral or written response. For guidelines about writing in teams, refer to Chapter 2.

Background for Assignment

Assume you and your team members comprise one of several teams from a private consulting firm. The firm has been hired to help plan a hotel/conference center to be built on your campus. Although the center will cater to some private clients, most customers will be associated with your institution—for example, parents of students, candidates for teaching or administrative positions, and participants in academic conferences.

Obviously, a project of this sort requires careful planning. One step in the process is to assess the needs of various people and groups that will occupy the center.

Following are listed just a few of the many groups or departments whose needs should be considered:

Accounting   Landscaping   Registration
Catering   Maintenance   Sales and Marketing
Computing   Procurement   Security
Housekeeping   Recreation   Training

The topics range broadly because the facility will have multiple purposes for a diverse audience.

Team Assignment

The consulting firm—of which your team is a part—will issue a joint report that describes the needs of all groups who will work in the new hotel/conference center. Assume that your team's task is to produce just a portion of the outline—not the text—of the report. Your outline will address one or more of the needs reflected in the previous list, or other needs of your choosing that have not been listed.

Coverage of International Communication

Because globalism continues to transform the business world, this book includes suggestions for understanding other cultures and for writing in an international context. In addition, each chapter's set of exercises ends with an "international communication assignment."

Assignments on Ethics

To reinforce the ethical guidelines described in Chapter 2, each chapter includes an ethics assignment. No one can escape the continuous stream of ethical decisions required of every professional almost every day, which is why assignments address ethical issues.

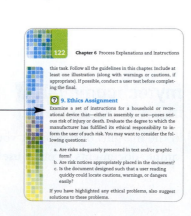

122   Chapter 6 Process Explanations and Instructions

this task. Follow all the guidelines in this chapter. Include at least one illustration (along with warnings or cautions, if appropriate). If possible, conduct a user test before completing the final.

9. Ethics Assignment

Examine a set of instructions for a household or recreational device that—either in assembly or use—poses serious risk of injury or death. Evaluate the degree to which the manufacturer has fulfilled its ethical responsibility to inform the user of such risk. You may want to consider the following questions:

a. Are risks adequately presented in text and/or graphic form?
b. Are risk notices appropriately placed in the document?
c. Is the document designed such that a user reading quickly could locate cautions, warnings, or dangers easily?

If you have highlighted any ethical problems, also suggest solutions to these problems.

Appendix A

>>> Handbook

This handbook includes entries on the basics of writing. It contains three main types of information:

1. **Grammar:** the rules by which we edit sentence elements. Examples include rules for the placement of punctuation, the agreement of subjects and verbs, and the placement of modifiers.
2. **Mechanics:** the rules by which we make final proofreading changes. Examples include the rules for abbreviations and the use of numbers. A list of commonly misspelled words is also included.
3. **Usage:** information on the correct use of particular words, especially pairs of words that are often confused. Examples include problem words like *affect/effect*, *complement/compliment*, and *who/whom*.

This handbook is presented in alphabetized fashion for easy reference during the editing process. Grammar and mechanics entries are in all uppercase; usage entries are in lowercase. Several exercises follow the entries.

A/An

A and *an* are different forms of the same article. *A* occurs before words that start with consonants or consonant sounds. EXAMPLES:

- a three-pronged plug
- a once-in-a-lifetime job (*once* begins with the consonant sound of *w*)
- a historic moment (many speakers and some writers mistakenly use *an* before *historic*)

An occurs before words that begin with vowels or vowel sounds. EXAMPLES:

- an eager new employee
- an hour before closing

A lot/Alot

The correct form is the two-word phrase *a lot*. Although acceptable in informal discourse, *a lot* usually should be replaced by more formal diction in technical writing. EXAMPLE: "They retrieved many [*not a lot of*] soil samples from the construction site."

243

Handbook

This book provides a well-indexed, alphabetized handbook on grammar, mechanics, and usage. Handbook gives quick access to rules for eliminating editing errors during the revision process.

Information on English as a Second Language

A growing number of technical communication students are from other countries or cultures where English is not the first language. The English as a Second Language (ESL) section of the Handbook focuses on three main problem areas: articles, prepositions, and verb use. It also applies ESL analysis to an excerpt from a technical report.

English as a Second Language (ESL)

Technical writing challenges native English speakers and nonnative speakers alike. The purpose of this section is to present a basic description of three grammatical forms: articles, verbs, and prepositions. These forms may require more intense consideration from international students when they complete technical writing assignments. Each issue is described using the ease-of-operation section from a memo about a fax machine. The passage, descriptions, and charts work together to show how these grammar issues function collectively to create meaning.

>>>Your One-Stop Source for *Technical Communication Resources*

MyTechCommLab for *Technical Communication Fundamentals*

MyTechcommlab contains a wealth of multimedia technical communication resources in one easy-to-use location. Resources include the following:

- **Over 90 Model Documents:** Most of these documents include interactive activities and annotations selected from a variety of professions and purposes (letters, memos, career correspondence, proposals, reports, instructions and procedures, descriptions and definitions, Web sites, and presentations). MyTechcommlab also contains **50 Interactive Documents** that include rollover annotations highlighting purpose, audience, design, and other critical topics.

- **Grammar, Mechanics, and Writing Help:** If students need more practice in basic grammar and usage, MyTechCommLab's grammar diagnostics will generate a study plan linked to the thousands of test items in ExerciseZone, with results tracked by Pearson's exclusive GradeTracker.

- **Document Design Resources:** A **Writing Process tutorial** leads students through each stage of the writing process—from prewriting to final formatting. A new **tutorial on Writing Formal Reports** offers step-by-step guidance for creating one of the most common document types in technical communication and working with sources. **Activities and Case Studies** provide over 65 exercises, all rooted in technical communication and many document-based, including three new case studies on usability. An **online reference library of e-books** includes pdf files for books on Visual Communication and Workplace Literacy.

- **Pearson's MySearchLab:** Research, Grammar, and Writing Tips access gives students research tips, access to the EBSCO document database, writing and assessment and instruction and access to the Longman Online Handbook for Writers.

- **E-book with Online Reference Sources:** Students can choose to purchase a version of MyTechcommlab that includes an e-book with links embedded in the pages to all online resources.

To preview MyTechCommLab, go to www.mytechcommlab.com.

If your textbook did not come packaged with an access code, standalone access codes with or without an integrated e-book can also be purchased online at www.pearsonhighered.com.

>>> Instructor's Resources

■ Instructor's Manual

An expanded Instructor's Manual, loaded with helpful teaching notes for your classroom, including answers to the chapter quiz questions, a test bank, and instructor notes for assignments and activities, is located on the Companion Web site.

■ Test Generator
■ PowerPoint Lecture Presentation Package

The Instructor's Manual, Test Generator, and PowerPoint Package can be downloaded from the Instructor's Resource Center. To access supplementary materials online, instructors need to request an instructor access code. Go to **www.pearsonhighered.com/irc**, where you can register for a code. Within 48 hours of registering you will receive a confirming e-mail, including an instructor access code. Once you have received your code, locate your text in the online catalog and click on the Instructor Resources button on the left side of the catalog product page. Select a supplement and a log-in page will appear. Once you have logged in, you can access instructor material for all Pearson textbooks.

>>> Acknowledgments

We would like to thank the following reviewers of the seventh edition of our original textbook for their helpful insights that are also included in this *Fundamentals* text:

Heidi Hatfield Edwards, *Florida Institute of Technology*

Liz Kleinfeld, *Red Rocks Community College*

Brian Van Horne, *Metropolitan State College of Denver*

In addition, the following reviewers have helped throughout the multiple editions of this book:

Brian Ballentine, *Case Western Reserve University*

Jay Goldberg, *Marquette University*

Linda Grace, *Southern Illinois University*

Darlene Hollon, *Northern Kentucky University*

John Puckett, *Oregon Institute of Technology*

Kirk Swortzel, *Mississippi State University*

Catharine Schauer, *Visiting Professor, Embry Riddle University*

Friends and colleagues who contributed to this edition and/or other editions include Shawn Tonner, Mark Stevens, Saul Carliner, George Ferguson, Alan Gabrielli, Bob Harbort, Mike Hughes, Dory Ingram, Becky Kelly, Chuck Keller, Monique Logan, Jo Lundy, Minoru Moriguchi, Randy Nipp, Jeff Orr, Ken Rainey, Lisa A. Rossbacher, Betty Oliver

Seabolt, Hattie Schumaker, John Sloan, Herb Smith, Lavern Smith, James Stephens, John Ulrich, Steven Vincent, and Tom Wiseman.

Four companies allowed us to use written material gathered during Sandy's consulting work: Fugro-McClelland, Law Engineering and Environmental Services, McBride-Ratcliff and Associates, and Westinghouse Environmental and Geotechnical Services. Though this book's fictional firm, M-Global, Inc., does have features of the world we observed as consultants, we want to emphasize that M-Global is truly an invention.

Sandy would like to thank the following students for allowing us to adapt their written work for use in this book: Michael Alban, Becky Austin, Corey Baird, Natalie Birnbaum, Cedric Bowden, Gregory Braxton, Ishmael Chigumira, Bill Darden, Jeffrey Daxon, Rob Duggan, William English, Joseph Fritz, Jon Guffey, Sam Harkness, Gary Harvey, Lee Harvey, Hammond Hill, Sudhir Kapoor, Steven Knapp, Wes Matthews, Kim Meyer, James Moore, Chris Owen, Scott Lewis, James Porter, James Roberts, Mort Rolleston, Chris Ruda, Barbara Serkedakis, Tom Skywark, Tom Smith, DaTonja Stanley, James Stephens, Chris Swift, and Jeff Woodward. Kaye would like to thank her research assistants, Rachel Stancliff and Ted Koehler, who identified outdated examples and references and provided updated references, examples, and models.

We want to give special thanks to our Pearson Education editor, Gary Bauer, for suggesting the concept for this book and to Rex Davidson, our production editor at Pearson.

Brief Contents

Contents

Chapter | **3** | **Visual Design   43**

Chapter | **4** | **Letters, Memos, and Electronic Communication   75**

Chapter | **5** | **Definitions and Descriptions   93**

Chapter | **6** | ## Process Explanations and Instructions   108

Chapter | **7** | ## Reports   123

Technical Communication Fundamentals

Chapter | 1 | # Introduction to Technical Communication

>>> Chapter Outline

Good communication skills are essential in any career you choose. Jobs, promotions, raises, and professional prestige result from your ability to present both written and visual information effectively.

>>> Defining Technical Communication

mytechcommlab

For review, see the Writing Process section.

You may have learned how to write academic essays in previous writing courses. This book helps you transfer that basic knowledge to the kind of writing done on the job. Career writing is so practical, so well-grounded in common sense, that it will seem to proceed smoothly from your previous work. While traditional academic writing and technical communication share many common features, they also have important differences.

Features of Academic Writing

Writing you have done in school probably has had the following characteristics:

- **Purpose:** Communicating what you know about the topic, in a way that justifies a high grade
- **Your knowledge of topic:** Less than the teacher who evaluates the writing
- **Audience:** The teacher who requests the assignment and reads it from beginning to end
- **Criteria for evaluation:** Depth, logic, clarity, unity, supporting evidence, and grammar
- **Graphic elements:** Sometimes used to explain and persuade

Academic writing requires that you use words to display your learning to someone who knows more about the subject than you do. Because this person's job is to evaluate your work, you have what might be called a *captive audience*. In an academic setting, the purpose is to demonstrate your command of information to someone more knowledgeable about the subject than you are.

Features of Technical Communication

The rules for writing shift when you begin your career. *Technical communication* is a generic term for all written and oral communication done on the job—whether in business, industry, or in other settings. It is particularly identified with documents in technology, engineering, science, the health professions, and other fields with specialized vocabularies. The terms *technical writing, professional writing, business writing,* and *occupational writing* also refer to writing done in your career.

mytechcommlab

My Tech Comm Lab includes examples of technical communication genres.

Besides writing projects, your career will also bring you speaking responsibilities, such as formal speeches at conferences and informal presentations at meetings. Thus, the term *technical communication* can encompass the full range of writing and speaking responsibilities required to communicate your ideas on the job. The following discusses the main characteristics of technical communication:

- **Purpose:** Getting something done within an organization or helping a customer, client, or colleague get something done
- **Your knowledge of topic:** Usually greater than that of the reader

■ **Audience:** Often several people with differing technical backgrounds

■ **Criteria for evaluation:** Clear and simple organization of ideas and supporting detail appropriate to the needs of busy readers

■ **Graphic elements:** Frequently used to explain existing conditions and to present alternative courses of action

Contrast these features with those of academic writing, listed earlier. In particular, note the following main differences:

1. Technical communication aims to help people make decisions and perform tasks, whereas academic writing aims only to display your knowledge.

2. Technical communication usually responds to the needs of the workplace, whereas academic writing usually responds to an assignment created by a teacher.

3. Technical communication is created by an informed writer conveying needed information both verbally and visually to an uninformed reader, whereas academic writing is created by a student as the learner for a teacher as the source of knowledge.

4. Technical communication often is read by many readers, whereas academic writing usually aims to satisfy only one person, the teacher.

Finally, technical communication places greater emphasis on techniques of organization and visual cues that help readers find important information as quickly as possible.

Figure 1-1 lists some typical on-the-job writing assignments. Although not exhaustive, the list does include many of the writing projects you will encounter. Figure 1-2 is an

■ **Figure 1-1** ■
Examples of technical communication

Correspondence: In-House or External
- Memos to your boss and to your subordinates
- Routine letters to customers, vendors, etc.
- "Good news" letters to customers
- "Bad news" letters to customers
- Sales letters to potential customers
- Electronic mail (e-mail) messages to coworkers or customers over a computer network

Short Reports: In-House or External
- Analysis of a problem
- Recommendation
- Equipment evaluation
- Progress report on project or routine periodic report
- Report on the results of laboratory or field work
- Description of the results of a company trip

Long Reports: In-House or External
- Complex problem analysis, recommendation, or equipment evaluation
- Project report on field or laboratory work
- Feasibility study

Other Examples
- Proposal to boss for new product line
- Proposal to boss for change in procedures
- Proposal to customer to sell a product, a service, or an idea
- Proposal to funding agency for support of research project
- Abstract or summary of technical article
- Technical article or presentation
- Operation manual or other manual
- Web site

■ **Figure 1-2** ■ Short
report

MEMORANDUM

DATE: December 6, 2011
TO: Holly Newsome
FROM: Michael Allen
SUBJECT: Printer Recommendation

Introductory Summary

Recently you asked for my evaluation of the Hemphill 5000 printer/fax/scanner/copier currently used in my department. Having analyzed the machine's features, print quality, and cost, I am quite satisfied with its performance.

Features

Among the Hemphill 5000's features, I have found these five to be the most useful:
1. Easy to use control panel
2. Print and copy speed of up to 34 pages per minute for color and black and white
3. Ability to print high-quality documents like brochures & report covers
4. Built-in networking capability
5. Ability to scan documents to or from a USB port

In addition, the Hemphill 5000 offers high-quality copies, color copies and faxes, and it uses high capacity ink cartridges to reduce costs.

Print Quality

The Hemphill 5000 produces excellent prints that rival professional typeset quality. The print resolution is 1200 x 1200 dots per inch, among the highest attainable in printer/fax/scanner/copier combinations. This memo was printed on the 5000, and, as you can see, the quality speaks for itself.

Cost

Considering the features and quality, the 5000 is an excellent network combination printer for workgroups within the firm. At a retail price of $239, it is also one of the lowest-priced combination printers, yet it comes with a two-year warranty and excellent customer support.

Conclusion

On the basis of my observation, I strongly recommend that our firm continue to use and purchase the Hemphill 5000. Please call me at ext. 204 if you want further information about this excellent machine.

M-Global Inc. | 127 Rainbow Lane | Baltimore MD 21202 | 410.555.8175

example of a short technical document. Note that it has the features of technical communication listed previously.

1. It is written to get something done—that is, to evaluate a printer

2. It is sent from someone more knowledgeable about the printer to someone who needs information about it

3. Although the memo is directed to one person, the reader probably will share it with others before making a decision concerning the writer's recommendation

4. It is organized clearly, moving from data to recommendations and including headings

5. It provides limited data to describe the features of the printer

Now that you know the nature and importance of technical communication, the next section examines the ethical context for communication done on the job.

>>> Determining the Purpose of Your Documents

If you have already taken a basic composition course, you will see similarities between rhetorical aims studied in that course and those in technical communication. Writing assignments you have had in school have probably asked you to *inform* your reader about an event or object, to *analyze* a process or idea, or to *argue* the strength or weakness of an interpretation or theory.

Information: When readers pick up a technical communication document, they may want to know how to perform an operation or follow an established procedure. They may want to make an informed decision. Clear, reliable information is the basis of analysis and argument.

Analysis: At first, it may not seem like analysis is an important purpose of workplace writing, but it is essential to problem solving and decision making. You may be asked to analyze options for a supervisor who will make a recommendation to a client or you may be asked to use analysis to make your own recommendation.

Argument: Good argument forms the basis for all technical communication. Some people have the mistaken impression that only recommendation reports and proposals argue their case to the reader, and that all other writing should be objective rather than argumentative. The fact is, every time you commit words to paper, you are arguing your point.

Every piece of your workplace writing should have a specific reason for being. The purpose may be dictated by someone else or selected by you. In either case, it must be firmly understood *before* you start writing. Purpose statements guide every decision you make while you plan, draft, and revise. Your choice of purpose statement will fall somewhere within this continuum:

Neutral, objective statement ⟷ Persuasive, subjective statement

For example, when reporting to your boss on the feasibility of adding a new wing to your office building, you should be quite objective. You must provide facts that can lead to an informed decision by someone else. If you are an outside contractor proposing to construct such a wing, however, your purpose is more persuasive. You will be trying to convince readers that your firm should receive the construction contract.

When preparing to write, therefore, ask yourself two related questions about your purpose.

>> **Question 1: Why Am I Writing This Document?**

This question should be answered in just one or two sentences, even in complicated projects. Often the resulting purpose statement can be moved as is to the beginning of your outline and later to the first draft.

>> **Question 2: What Response Do I Want from Readers?**

The first question about purpose leads inevitably to the second about results. Again, your response should be only one or two sentences long. Although brief, it should pinpoint exactly what you want to happen as a result of your document. Are you just giving data for the file? Will information you provide help others do their jobs? Will your document recommend a major change? Unlike the purpose statement, the results statement may not go directly into your document.

The answers to these two questions about purpose and results should be included on the Planning Form your instructor may ask you to use for assignments. Figure 1-3 on pages 7 and 8 includes a copy of the form along with instructions for using it. The last page of this book contains another copy you can duplicate for use with assignments.

Having established your purpose, you are now ready to consider the next part of the writing process: audience analysis.

>>> Analyzing Your Readers

One cardinal rule governs all on-the-job writing:

Write for your reader, not for yourself.

This rule especially applies to science and technology because many readers may know little about your field. The key to avoiding this problem is to examine the main obstacles readers face and adopt a strategy for overcoming them.

Obstacles for Readers

Readers of all backgrounds often have these four problems when reading any technical document:

1. Constant interruptions
2. Impatience finding information they need
3. A different technical background from the writer
4. Shared decision-making authority with others

If you think about these obstacles every time you write, you will be better able to understand and respond to your readers.

>> **Obstacle 1: Readers Are Always Interrupted**

As a professional, how often will you have the chance to read a report or other document without interruption? Such times are rare. Your reading time will be interrupted by

PLANNING FORM

Name: _____ Assignment: _____

I. Purpose: Answer each question in one or two sentences.

    A. Why are you writing this document?

    B. What response do you want from readers?

II. Audience

    A. Reader Matrix: Fill in names and positions of people who may read the document

	Decision Makers	Advisers	Receivers
Managers			
Experts			
Operators			
General Readers			

    B. Information on individual readers: Answer these questions about the primary audience for this document. If the primary audience includes more than one reader (or type of reader) and there are significant differences between the readers, answer the questions for each (type of) reader. Attach additional sheets as necessary.

Primary audience:

1. What is this reader's technical or educational background?

2. What main question does this person need answered?

3. What main action do you want this person to take?

4. What features of this person's personality might affect his or her reading?

III. Document

    A. What information do I need to include in the

        1. Abstract?

        2. Body?

        3. Conclusion?

    B. What organizational patterns are appropriate to the subject and purpose?

    C. What style choices will present a professional image for me and the organization I represent?

■ **Figure 1-3** ■ Planning Form for all technical documents

Instructions for Completing the Planning Form

The Planning Form is for your use in preparing assignments in your technical communication course. It focuses only on the planning stage of writing. Complete it before you begin your first draft.

1. Use the Planning Form to help plan your strategy for all writing assignments. Your instructor may or may not require that it be submitted with assignments.

2. Photocopy the form on the back page of this book or write the answers to questions on separate sheets of paper, whatever option your instructor prefers. (Your instructor may ask you to use an electronic version or enlarged, letter-sized copies of the form that are included in the Instructor's Resource Manual.)

3. Answer the two purpose questions in one or two sentences each. Be as specific as possible about the purpose of the documents and the response you want—especially from the decision makers.

4. Note that the reader matrix classifies each reader by two criteria: (a) technical levels (shown on the vertical axis) and (b) relationship to the decision-making process (shown on the horizontal axis). Some of the boxes will be filled with one or more names whereas others may be blank. How you fill out the form depends on the complexity of your audience and, of course, on the directions of your instructor.

5. Note that the "Information on Individual Readers" section can be filled out for one or more readers, depending on what your instructor requires.

6. Answer the document questions in one or two sentences each. Refer to Chapter 1 for information about the ABC format and organizing patterns that can be used in documents.

■ **Figure 1-3** ■ Continued

meetings and phone calls, so a report often gets read in several sittings. Aggravating this problem is the fact that readers may have forgotten details of the project.

>> Obstacle 2: Readers Are Impatient

Many readers lose patience with vague or unorganized writing. They think, "What's the point?" or "So what?" as they plod through memos, letters, reports, and proposals. They want to know the significance of the document right away.

>> Obstacle 3: Readers Lack Your Technical Knowledge

In college courses, the readers of your writing are professors who usually have knowledge of the subject on which you are writing. In your career, however, you will write to readers who lack the information and background you have. They expect a technically sophisticated response, but in language they can understand. If you write over their heads, you will not accomplish your purpose. Think of yourself as an educator; if readers do not learn from your reports, you have failed in your objective.

>> Obstacle 4: Most Documents Have More Than One Reader

Readers usually share decision-making authority with others who may read all or just part of the text. Thus, you must respond to the needs of many individuals—most of them have a hectic schedule, are impatient, and have a technical background different from yours.

Ways to Understand Readers

Obstacles to communication can be frustrating, yet there are techniques for overcoming them. First, you must try to find out exactly what information each reader needs. Think of the problem this way—would you give a speech without learning about the background of your audience? Writing depends just as much, if not more, on such analysis. Follow these four steps to determine your readers' needs:

mytechcommlab
See Model Report 11: Position Paper for a model document that clearly addresses its purpose and audience.

>> Audience Analysis Step 1: Write Down What You Know About Your Reader

To build a framework for analyzing your audience, you need to write down—not just casually think about—the answers to these questions for each reader:

1. What is this reader's technical or educational background?
2. What main question does this person need the answer to?
3. What main action do you want this person to take?
4. What features of this person's personality might affect his or her reading?

   The Planning Form in Figure 1-3 includes these four questions.

>> Audience Analysis Step 2: Talk With Colleagues Who Have Written to the Same Readers

Often your best source of information about readers is a colleague where you work. Ask around the office or check company files to discover who else may have written to the same audience. Useful information could be as close as the next office.

>> Audience Analysis Step 3: Find Out Who Makes Decisions

Almost every document requires action of some kind. Identify decision makers ahead of time so that you can design the document with them in mind. Know the needs of your *most important* reader.

>> **Audience Analysis Step 4:** Remember That All Readers Prefer Simplicity

Even if you uncover little specific information about your readers you can always rely on one basic fact: Readers of all technical backgrounds prefer concise and simple writing.

Types of Readers

To complete the audience-analysis stage, this section shows you how to classify readers by two main criteria: knowledge and influence. Specifically, you must answer two questions about every potential reader:

1. How much does this reader already know about the subject?
2. What part will this reader play in making decisions?

Then use the answers to these questions to plan your document. The Planning Form in Figure 1-3 provides a reader matrix by which you can quickly view the technical levels and decision-making roles of all your readers. For complex documents, your audience may include many of the 12 categories shown on the matrix. Also, you may have more than one person in each box—that is, there may be more than one reader with the same background and decision-making role.

Technical Levels

On-the-job writing requires that you translate technical ideas into language that nontechnical people can understand. This task can be very complicated because you often have several readers, each with different levels of knowledge about the topic. If you are to "write for your reader, not for yourself," you must identify the technical background of each reader. Four categories help you classify each reader's knowledge of the topic.

>> Reader Group 1: Managers

Many technical professionals aspire to become managers. Once into management, they may be removed from hands-on technical details of their profession. Instead, they manage people, set budgets, and make decisions of all kinds. Thus you should assume that management readers are not familiar with fine technical points, have forgotten details of your project, or both. These managers often need

■ Background information
■ Definitions of technical terms
■ Lists and other format devices that highlight points
■ Clear statements about what is supposed to happen next

>> Reader Group 2: Experts

Experts include anyone with a good understanding of your topic. They may be well-educated—as with engineers and scientists—but that is not necessarily the case. Whatever their educational levels, most experts in your audience need

- Thorough explanations of technical details
- Data placed in tables and figures
- References to outside sources used in writing the report
- Clearly labeled appendices for supporting information

>> Reader Group 3: Operators

Because decision makers are often managers or technical experts, these two groups tend to get most of the attention. However, many documents also have readers who are operators. They may be technicians in a field crew, workers on an assembly line, salespeople in a department store, or drivers for a trucking firm—anyone who puts the ideas in your document into practice. These readers expect

- A clear table of contents for locating sections that relate to them
- Easy-to-read listings for procedures or instructions
- Definitions of technical terms
- A clear statement of exactly how the document affects their jobs

>> Reader Group 4: General Readers

General readers often have the least amount of information about your topic or field. For example, a report on the environmental impact of a toxic waste dump might be read by general readers who are homeowners in the surrounding area. Most will have little technical understanding of toxic wastes and associated environmental hazards. These general readers often need

- Definitions of technical terms
- Frequent use of graphics, such as charts and photographs
- A clear distinction between facts and opinions

As with managers, general readers must be assured that (1) all implications of the document have been put down on paper and (2) important information has not been buried in overly technical language.

Decision-Making Levels

Figure 1-3 shows that your readers, whatever their technical level, can also be classified by the degree to which they make decisions based on your document. Use the following three levels to classify your audience during the planning process:

>> First-Level Audience: Decision Makers

The first-level audience, the *decision makers,* must act on the information. If, for example, you are comparing two computer systems for storing records at a hospital, the first-level audience decides which unit to purchase. In other words, decision makers translate information into action. They are usually, but not always, managers within the organization.

>> Second-Level Audience: Advisers

This second group could be called *influencers*. Although they do not make decisions themselves, they read the document and give advice to those who make the decisions. Often, the second-level audience is composed of experts, such as engineers and accountants, who are asked to comment on technical matters. After reading the summary, a decision-making manager may refer the rest of the document to advisers for their comments.

>> Third-Level Audience: Receivers

Some readers do not take part in the decision-making process, but only receive information contained in the document. For example, a report recommending changes in the hiring of fast-food workers may go to the store managers after it has been approved, just so they can put the changes into effect. This third-level audience usually includes readers defined as *operators* in the previous section—that is, those who may be asked to follow guidelines or instructions contained in a report.

Using all this information about technical and decision-making levels, you can analyze each reader's (1) technical background with respect to your document and (2) potential for making decisions after reading what you present. Then you can move on to the research and outline stages of writing.

>>> Planning Your Documents

Given the varied backgrounds and interests of your readers, you must answer one essential question: How can you best organize information to satisfy so many different people?

Figure 1-4 shows you three possible options for organizing information for the technical expertise of a mixed technical audience, but only one is recommended in this book. Writers who choose Option A direct their writing to the most technical people. Writers

■ Figure 1-4 ■
Options for organizing information

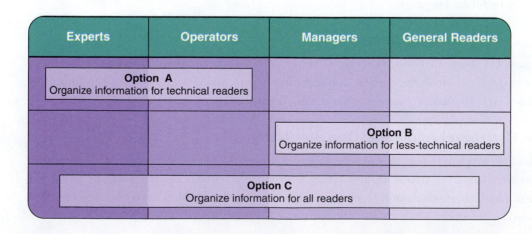

Experts	Operators	Managers	General Readers
Option A Organize information for technical readers			
		Option B Organize information for less-technical readers	
Option C Organize information for all readers			

who choose Option B respond to the dilemma of a mixed technical audience by finding the lowest common denominator—that is, they write to the level of the least technical person. Each option satisfies one segment of readers at the expense of the others.

Option C is preferred in technical communication for mixed readers. It encourages you to organize documents so that all readers—both technical and nontechnical—get what they need. The rest of this chapter provides strategies for developing this option. It describes general principles of organization and guidelines for organizing entire documents, individual document sections, and paragraphs.

Three Principles of Organization

Good organization starts with careful analysis of your audience. Most readers are busy, and they skip around as they read. Think about how you examine a news organization's Web site or a weekly newsmagazine. You are likely to take a quick look at articles of special interest to you; then you might read them more thoroughly, if there is time. That approach also resembles how your audience treats technical reports and other work-related documents. If important points are buried in long paragraphs or sections, busy readers may miss them. Three principles respond realistically to the needs of your readers:

>> Principle 1: Write Different Parts for Different Readers

The longer the document, the less likely it is that anyone will read it from beginning to end. As shown in Figure 1-5, they use a *speed-read approach* that includes these steps:

Step 1: **Quick scan.** Readers scan easy-to-read sections like executive summaries, introductory summaries, introductions, tables of contents, conclusions, and recommendations. They pay special attention to beginning and end sections, especially in documents longer than a page or two, and to illustrations.

Step 2: **Focused search.** Readers go directly to parts of the document body that give them what they need at the moment. To find information quickly, they search for navigation devices like subheadings, listings, and white space in margins to guide their reading. (See Chapter 3 for a discussion of page design.)

Step 3: **Short follow-ups.** Readers return to the document, when time permits, to read or reread important sections.

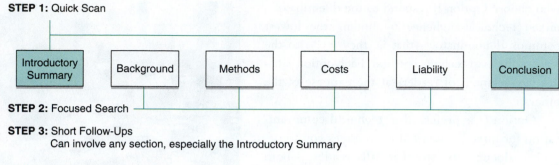

STEP 1: Quick Scan

| Introductory Summary | Background | Methods | Costs | Liability | Conclusion |

STEP 2: Focused Search

STEP 3: Short Follow-Ups
   Can involve any section, especially the Introductory Summary

■ **Figure 1-5** ■ Sample speed-read approach to short proposal

Your job is to write in a way that responds to this nonlinear and episodic reading process of your audience. Most important, you should direct each section to those in the audience most likely to read that particular section.

>> Principle 2: Emphasize Beginnings and Endings

Suspense fiction relies on the interest and patience of readers to piece together important information. The writer usually drops hints throughout the narrative before finally revealing who did what to whom. Technical communication operates differently. Busy readers expect to find information in predictable locations without having to search for it. Their first-choice locations for important information are as follows:

■ The beginning of the entire document
■ The beginnings of report sections
■ The beginnings of paragraphs

Emphasizing beginnings and endings responds to the reading habits and psychological needs of readers. At the beginning, they want to know where you are heading. They need a simple road map for the rest of the passage. In fact, if you do not provide something important at the beginnings of paragraphs, sections, and documents, readers will start guessing the main point themselves. At the ending, readers expect some sort of wrap-up or transition; your writing should not simply drop off. The following paragraph begins and ends with such information (italics added):

> *Already depleted sea turtle, marine mammal, seabird, and noncommercial fish populations are endangered by incidental capture in fishing gear.* Worldwide, about 25 percent of the catch is discarded, either because it is not commercially valuable or because of regulatory requirements that prohibit keeping undersized or nontargeted marine life. Destructive fishing practices, such as bottom trawling and dredging, are damaging vital habitat upon which fish and other living resources depend. *Taken together, overfishing, bycatch, and habitat destruction are changing relationships among species in food webs and altering the function of marine ecosystems.* [Pew Oceans Commission, America's Living Oceans: Charting a Course for Sea Change (Arlington, VA: Pew Oceans Commission, 2003), 5, 9.]

The first sentence gives readers an immediate impression of the two topics to be covered in the paragraph. The paragraph body explores details of both topics. Then the last sentence flows smoothly from the paragraph body by reinforcing the main point about over fishing.

>> Principle 3: Repeat Key Points

You have learned that different people focus on different sections of a document. Sometimes no one carefully reads the entire report. For example, managers may have time to read only the summary, whereas technical experts may skip the leadoff sections and go directly to "meaty" technical sections with supporting information. These varied reading patterns require a *redundant* approach to organization—you must repeat important information in different sections for different readers. Your strategic repetition of a major finding, conclusion, or recommendation gives helpful reinforcement to readers always searching for an answer to the "So what?" question as they read. Now we are ready to be more specific about how the three general principles of organization apply to documents, document sections, and paragraphs.

ABC Format for Documents

Technical documents should assume a three-part structure that consists of a beginning, a middle, and an end. This book labels this structure the *ABC Format* (for *A*bstract, *B*ody, and *C*onclusion). Visually, think of this pattern as a three-part diamond structure, as shown in Figure 1-6:

- **Abstract:** A brief beginning component is represented by the narrow top of the diamond, which leads into the body.
- **Body:** The longer middle component is represented by the broad, expansive portion of the diamond figure.
- **Conclusion:** A brief ending component is represented by the narrow bottom of the diamond, which leads away from the body.

First and foremost, the ABC format pertains to the organization of entire documents. However, the same beginning-middle-end strategy applies to the smaller units of discourse—document sections and paragraphs.

Figure 1-2 (page 4) includes a memo report that conforms to this structure. The following sections discuss the three ABC components in detail.

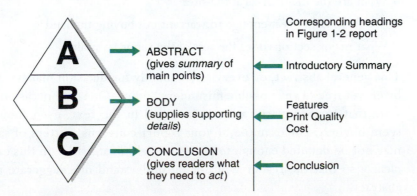

■ **Figure 1-6** ■ ABC format for all documents

Document Abstract: The "Big Picture" for Decision Makers

Every document should begin with an overview. As used in this text, *abstract* is defined as follows:

Abstract: brief summary of a document's main points. Although its makeup varies with the type and length of the document, an abstract usually includes (1) a clear purpose statement for the document, (2) the most important points for decision makers, and (3) a list or description of the main sections that follow the abstract. As a capsule version of the entire document, the abstract should answer readers' typical mental questions, such as: "How does this document concern me?" "What's the bottom line?" "So what?"

> **mytechcommlab**
>
> For more examples, see Model Abstract 1: Report and Model Abstract 2: Proposal.

Abstract information is provided under different headings, depending on the document's length and degree of formality. Some common headings are "Summary," "Executive Summary," "Introductory Summary," "Overview," and "Introduction." The abstract may vary in length from a short paragraph to a page or so. Its purpose, however, is always the same: to provide decision makers with highlights of the document.

For example, assume you are an engineer who has evaluated environmental hazards for the potential purchaser of a shopping mall site. Here is how the summary might read:

> As you requested, we have examined the possibility of environmental contamination at the site being considered for the new Klinesburg Mall. Our field exploration revealed two locations with deposits of household trash, which can be easily cleaned up. Another spot has a more serious deposit problem of 10 barrels of industrial waste. However, our inspection of the containers and soil tests revealed no leaks.
>
> Given these limited observations and tests, we conclude that the site poses no major environmental risks and recommend development of the mall. The rest of this report details our field activities, test analyses, conclusions, and recommendations.

You have provided the reader with a purpose for the report, an overview of important information for decision makers, and a reference to the four sections that follow. In doing so, you have answered the following questions, among others, for the readers:

- What are the major risks at the site?
- Are these risks great enough to warrant not buying the land?
- What major sections does the rest of the report contain?

This general abstract, or overview, is mainly for decision makers. Highlights must be brief, yet free of any possible misunderstanding. On some occasions, you may need to state that further clarification is included in the text, even though that point may seem obvious. For example, if your report concerns matters of safety, the overview may not be detailed enough to prevent or eliminate risks. In this case, state this point clearly so that the reader will not misunderstand or exaggerate the purpose of the abstract.

Document Body: Details for All Readers

The longest part of any document is the body. As used in this book, the *body* is defined as follows:

Body: the middle section(s) of the document providing supporting information to readers, especially those with a technical background. Unlike the abstract and conclusion, the body component allows you to write expansively about items, such as (1) the background of the project; (2) field, lab, office, or any other work on which the document is based; and (3) details of any conclusions, recommendations, or proposals that might be highlighted at the beginning or end of the document. The body answers this main reader question: "What support is there for points put forth in the abstract at the beginning of the document?"

Managers may read much of the body, especially if they have a technical background and if the document is short. Yet the more likely readers are technical specialists who (1) verify technical information for the decision makers or (2) use your document to do their jobs. In writing the body, use the following guidelines:

- **Separate fact from opinion.** Never leave the reader confused about where opinions begin and end. Body sections usually move from facts to opinions that are based on facts. To make the distinction clear, preface opinions with phrases, such as "We believe that," "I feel that," "It is our opinion that," and the like. Such wording gives a clear signal to readers that you are presenting judgments, conclusions, and other nonfactual statements. Also, you can reinforce the facts by including data in graphics.

- **Adopt a format that reveals the structure.** Use frequent headings and subheadings to help busy readers locate important information immediately.

- **Use graphics whenever possible.** Use graphics to draw attention to important points. Today more than ever, readers expect visual reinforcement of your text, particularly in more persuasive documents like proposals.

By following these guidelines, which apply to any document, you will make detailed body sections as readable as possible. They keep ideas from becoming buried in text and show readers what to do with the information they find.

Document Conclusion: Wrap-Up Leading to Next Step

Your conclusion deserves special attention, for readers often recall first what they have read last. We define the *conclusion* component as follows:

Conclusion: the final section(s) of the document bringing readers—especially decision makers—back to one or more central points already mentioned in the body. Occasionally, it may include one or more points not previously mentioned. In any case, the conclusion provides closure to the document and often leads to the next step in the writer's relationship with the reader.

The conclusion component may have any one of several headings, depending on the type and length of the document. Possibilities include "Conclusion," "Closing," "Closing

Remarks," and "Conclusions and Recommendations." A conclusion component answers the following types of questions:

- What major points have you made?
- What problem have you tried to solve?
- What should the reader do next?
- What will you do next?
- What single idea do you want to leave with the reader?

Whichever alternative you choose, your goal is to return to the main concerns of the most important readers—decision makers. Both the abstract and conclusion, in slightly different ways, should respond to the needs of this primary audience.

>>> Ethics in the Workplace

This section presents a set of ethical guidelines for the workplace, and shows how ethical guidelines, when applied to writing, can have legal applications. Throughout this book are assignments in which your own ethical decisions play an important role.

Ethical Guidelines for Work

As with your personal life, your professional life holds many opportunities for demonstrating your views of what is right or wrong. Most occur daily and without much fanfare, but cumulatively they compose our personal approach to morality. Thus, our belief systems are revealed by the manner in which we respond to this continuous barrage of ethical dilemmas.

Obviously, not everyone in the same organization or profession has the same ethical beliefs, nor should they. After all, each person's understanding of right and wrong flows from individual experiences, upbringing, religious beliefs, and cultural values. Some *ethical relativists* even argue that ethics only makes sense as a descriptive study of what people do believe, not a prescriptive study of what they should believe. Yet there are some basic guidelines that, in our view, should be part of the decision-making process in every organization. Although they may be displayed in different ways in different cultures, they should transcend national identity, cultural background, and family beliefs. In other words, these guidelines represent what, ideally, should be the *core* values for employees at international companies.

The guidelines in this section are common in many professional codes of ethics. These are general guidelines and provide a good foundation for ethical behavior in the workplace. However, you should also become familiar with the ethical guidelines specific to your employer and professional organizations.

>> Ethics Guideline 1: Be Honest

First, you should relate information accurately and on time—to your colleagues, to customers, and to outside parties, such as government regulators. This guideline also means you should not mislead listeners or readers by leaving out important information that relates to a situation, product, or service, including information about any conflicts of interest. You should interpret data carefully and present estimates as accurately as possible. In other words, give those with whom you communicate the same information that you would want presented to you.

>> Ethics Guideline 2: Do No Harm

Technical communicators often work in fields that affect public health and safety. You should avoid practices, inaccuracies, or mistakes that can harm people or property.

You should also support a positive and constructive work atmosphere. One way to achieve such a working environment is to avoid words or actions calculated to harm others. For example, avoid negative, rumor-laden conversations that hurt feelings, spread unsupported information, or waste time.

>> Ethics Guideline 3: Be Fair

You should treat those around you fairly, regardless of differences in race, religion, disability, age, or gender. You should also be aware of, and respect, differences in culture. This is especially important as business becomes more global.

>> Ethics Guideline 4: Honor Intellectual Property Rights

Of course you should follow copyright and patent laws, but you should also respect the work that others have put into developing and presenting their ideas. Credit others for ideas, text, or images that you have used.

When collaborating with others, show appreciation for their contribution, and welcome their input. Offer and accept feedback that will make the final product stronger.

>> Ethics Guideline 5: Respect Confidentiality

Remember that you are acting on the part of both your employer and your clients. Disclose sensitive information only with permission, and obtain written releases before you share materials. This is especially important for contract and freelance workers, who must have a portfolio of accomplishments to share with prospective clients. If you share confidential information with a prospective client, you show that you can not be trusted with sensitive material.

>> Ethics Guideline 6: Be Professional

When you are working, you represent your profession. Not only does this mean that you should act in an honorable manner, but also that you should meet deadlines with quality work. One way to do this is to keep current on developments in your field. Join a professional organization like the Association of Computing Machinery's Special Interest Group on the Design of Communication (ACM SIGDOC), read journal and magazine articles in your field, and participate in continuing education activities.

In your career, you should develop and apply your own code of ethics, making sure that it follows these guidelines and the guidelines of your employer and your professional organization.

Ethics and Legal Issues in Writing

Writing—whether on paper, audiotape, videotape, or computer screen—presents a special ethical challenge for demonstrating your personal code of ethics. Some countries, such as the United States, have a fairly well-developed legal context for writing, which means you must pay great attention to detail as you apply ethical principles to the writing process. This section highlights some common guidelines.

■ **Acknowledge Sources for Information Other Than Common Knowledge** You are obligated to provide sources for any information other than common knowledge. *Common knowledge* is usually considered to be factual and non-judgmental information that could be found in general sources about a subject. The sources for any other types of information beyond this definition must be cited in your document.

■ **Seek Written Permission Before Borrowing Extensive Text** Generally, it is best to seek written permission for borrowing more than a few hundred words from a source, especially if the purpose of your document is profit. This so-called "fair use" is, unfortunately, not clearly defined and subject to varying interpretations. It is best to (1) consult a reference librarian or other expert for an up-to-date interpretation of the application of fair use to your situation and (2) err on the side of conservatism by asking permission to use information, if you have any doubt. This probably has not been an issue in papers you have written for school, because they were for educational use and were not going to be published. However, this issue should be addressed in any writing you do outside of school.

■ **Seek Written Permission Before Borrowing Graphics** Again, you probably have not been concerned about this issue in projects you have created in school, but you must seek permission for any graphics you borrow for projects created outside of school. This guideline applies to any nontextual element, whether it is borrowed directly from the original or adapted by you from the source. Even if the graphic is not copyrighted, such as in an annual report from a city or county, you should seek permission for its use.

■ **Seek Legal Advice When You Cannot Resolve Complex Questions** Some questions, such as the use of trademarks and copyright, fall far outside the expertise of most of us. In such cases it is best to consult an attorney who specializes in such law. Remember that the phrase *Ignorance is bliss* has led many a writer into problems that could have been prevented by seeking advice when it was relatively cheap—at the beginning. Concerning U.S. copyrights in particular, you might first want to consult free information provided by the U.S. Copyright Office at its Web site (www.copyright.gov).

In the final analysis, acting ethically on the job means thinking constantly about the way in which people are influenced by what you do, say, and write. Also, remember that what you write could have a very long shelf life, perhaps to be used later as a reference for legal proceedings. Always write as if your professional reputation could depend on it, because it just might.

>>> Chapter Summary

Technical communication refers to the many kinds of writing and speaking you will do in your career. In contrast to most academic writing, technical communication aims to get something done (not just to demonstrate knowledge), relays information from someone (you) more knowledgeable about the topic to someone (the reader) less knowledgeable about it, is read by people from mixed technical and decision-making levels, presents ideas clearly and simply, and often uses data and graphics to provide support.

Good technical communication calls on special skills, especially in organization. Writers should follow three guidelines for organizing information: (1) Write different parts of the document for different readers, (2) place important information at the beginnings and endings, and (3) repeat key points throughout the document.

This chapter recommends the ABC format for organizing technical documents. This format includes an *Abstract* (summary), a *Body* (supporting details), and a *Conclusion* (wrap-up and transition to next step). The abstract section is particularly important because most readers give special attention to the start of a document.

Your technical communication always exists within an ethical context. Use the general ethics guidelines presented in this chapter, but also become familiar with the ethical guidelines of your field of professional practice.

>>> Learning Portfolio

Collaboration at Work Outline for a Consulting Report

General Instructions

Each Collaboration at Work exercise applies strategies for working in teams to chapter topics. The exercise assumes you (1) have been divided into teams of about three to six students, (2) use team time inside or outside of class to complete the case, and (3) produce an oral or written response. For guidelines about writing in teams, refer to Chapter 2.

Background for Assignment

Assume you and your team members comprise one of several teams from a private consulting firm. The firm has been hired to help plan a hotel/conference center to be built on your campus. Although the center will cater to some private clients, most customers will be associated with your institution—for example, parents of students, candidates for teaching or administrative positions, and participants in academic conferences.

Obviously, a project of this sort requires careful planning. One step in the process is to assess the needs of various people and groups that will occupy the center.

Following are listed just a few of the many groups or departments whose needs should be considered:

Accounting	Landscaping	Registration
Catering	Maintenance	Sales and Marketing
Computing	Procurement	Security
Housekeeping	Recreation	Training

The topics range broadly because the facility will have multiple purposes for a diverse audience.

Team Assignment

The consulting firm—of which your team is a part—will issue a joint report that describes the needs of all groups who will work in the new hotel/conference center. Assume that your team's task is to produce just a portion of the outline—not the text—of the report. Your outline will address one or more of the needs reflected in the previous list, or other needs of your choosing that have not been listed.

Assignments

Your instructor will indicate whether the end-of-chapter Assignments should serve as the basis for class discussion, for written exercises, or for both.

1. Features of Technical Communication

Option A. Locate an example of technical communication (such as a user's guide, owner's manual, or a document borrowed from a family member or an acquaintance who works in a technical profession) and prepare a brief analysis in which you explain (1) the purpose for which the piece was written, (2) the apparent readers and their needs, (3) the way in which the example differs from typical academic writing, and (4) the relative success with which the piece satisfies this chapter's guidelines.

Option B. Using the following brief example of technical writing, prepare the analysis requested in Option A.

DATE: June 15, 2011
TO: Pat Jones
Office Coordinator
FROM: Sean Parker
SUBJECT: New Productivity Software

Introductory Summary

As you requested, I have examined the FreeWork open source productivity suite software we are considering. On the basis of my observations, I recommend we secure one copy of FreeWork and test it in our office for two months. Then after comparing it with the other two packages we have tested, we can choose one of the three productivity packages to use throughout the office.

Features of FreeWork

As we agreed, my quick survey of FreeWork involved reading the user's manual, completing the orientation disk, and reviewing installation options. Here are the five features of the package that seemed most relevant to our needs:

1. **Formatting Flexibility:** FreeWork includes diverse "style sheets" to meet our needs in producing reports, proposals, letters, memos, articles, and even brochures. By engaging just one command on the keyboard, the user can change style sheets—whereby the program will automatically place text in a specified format.
2. **Mailers:** For large mailings, we can take advantage of FreeWork's "Mail Out" feature that automatically places names from mailing lists on form letters.
3. **Documentation:** To accommodate our staff's research needs, FreeWork has the capacity to renumber and rearrange footnotes as text is being edited.
4. **Page Review:** This package's "PagePeek" feature permits the user to view an entire written page on the screen. Without having to print the document, he or she can then see how every page of text will actually look on the page.
5. **Tables of Contents:** FreeWork can create and insert page numbers on tables of contents, created from the headings and subheadings in the text.
6. **Spreadsheets:** FreeWork includes a powerful spreadsheet that can be integrated into documents.
7. **Database:** FreeWork's database component can create forms and reports that can be integrated into documents.
8. **Graphics:** FreeWork includes a basic drawing program that will probably meet our needs.

Conclusion

Though I gave FreeWork only a brief look, my survey suggests that it may be a strong contender for use in our office. If you wish to move to the next step of starting a two-month office test, just let me know.

2. Purpose and Audience

The following examples deal with the same topic in four different ways. Using this chapter's guidelines on purpose and audience, determine the main reason for which each excerpt was written and the technical level of the intended readers.

A. You can determine the magnitude of current flowing through a resistor by use of this process:

- Connect the circuit (power supply, resistor, ammeter, voltmeter).
- Set the resistor knob to a setting of "1."
- Turn the voltage adjusting knob to the left until it stops rotating.
- Switch the voltmeter to "On" and make sure it reads "0.00 volts."
- Switch the power supply to "On."
- Slowly increase the voltage on the voltmeter from 0 to 10 volts.
- Take the reading from the ammeter to determine the amount of current flowing through the resistor.

B. After careful evaluation of several testers, I strongly recommend that Langston Electronics Institute purchase 100 Mantra Multitesters for use in our laboratories in Buffalo, Albany, and Syracuse.

C. Selected specifications for the Ames Multitester are as follows:

- Rangers 43
- DC Voltage.............0–125–250mV 1.25–2.5–10–5–125–500–1000V
- AC Voltage 0–5–25–125–250–500–1000V
- DC Current 0–25–50µA–2.5–5–25–50–250–500mA–10A
- Resistance ....0–2K–20K–200K–20 Mega ohms
- Decibels ....–20 to +62 in db 8 ranges
- Accuracy ....±3% on DC measurements
  ±4% on AC measurements
  ±3% on scale length on resistance
- Batteries................one type AA penlight cell
- Fuse ....0.75A at 250V

Note that the accuracy rate for the Ames is within our requirements of ±6% and is considerably lower than the three other types of testers currently used by our staff.

D. Having used the Ames Multitester in my own home laboratory for the last few months, I found it extremely reliable during every experiment. In addition, it is quite simple to operate and includes clear instructions. As a demonstration of this operational ease, my 10-year-old son was able to follow the instructions that came with the device to set up a functioning circuit.

3. Audience Analysis

Find a commercial Web site (a Web site from a manufacturer or retailer) designed for children. Sites that promote cereal, toys, or snack foods are good choices.

- Is the Web site designed to inform, provide analysis, or to persuade? How do you know?
- What have the designers of the Web site done to appeal to their audience? What do their choices tell you about the results of their audience analysis?
- Is there a section on the Web site specifically targeted to parents? How does it differ from the Web pages for children? How is it similar to the pages for children?

4. Contrasting Audiences

Photocopy three articles from the same Sunday issue of a local or national newspaper. Choose each article from a different section of the paper—for example, you could use the sections on automobiles, business, travel, personal computers, national political events, local events, arts, editorials, or employment. Describe the intended audience of each article.

Then explain why you think the author has been successful or unsuccessful in reaching the particular audience for each article.

5. Rewrite for Different Audiencexy

Locate an excerpt from a technical article or textbook, preferably on a topic that interests you because of your background or college major. Rewrite all or part of the selection so that it can be understood by readers who have no previous knowledge of the topic.

6. Collecting and Organizing Information

Most word processing programs include a feature that allows users to track the changes that are made to documents, as well as insert questions, comments and advice for revision. This feature is especially useful for collaborative projects because it allows team members to see who recommended various changes, as well as allowing several people to comment on drafts.

A. Identify the reviewing features that are available in your word processing program and how they are accessed. Create a list of the features that you believe would be most useful for students working on team projects.

B. Organize the information you have collected into a single page reference for students who want to use the reviewing features in your word processing program.

7. Collecting and Organizing Information

Many Web sites offer advice to incoming freshmen about what to pack for their college dorm room. Using at least two lists as a starting place, create your own list of recommendations. You may include as many or as few of the recommended items as you feel worthwhile, and you can add your own items to the list. Then choose a principle for organizing the items on your list. Group the items and clearly identify the characteristics that helped you group the items. When you turn in your lists or share them with the class (as your teacher instructs), identify the Internet sites that you used as a starting point.

8. Writing an Abstract

The short report that follows lacks an abstract that states the purpose and provides the main conclusion or recommendation from the body of the report. Write a brief abstract for this report.

DATE: June 13, 2011
TO: Ed Simpson
FROM: Jeff Radner
SUBJECT: Creation of an Operator Preventive Maintenance Program

The Problem
The lack of operator involvement in the equipment maintenance program has caused the reliability of equipment to decline. Here are a few examples:

◆ A tractor was operated without adequate oil in the crankcase, resulting in a $15,000 repair bill after the engine locked.
◆ Operators have received fines from police officers because safety lights were not operating. The bulbs were burned out and had not been replaced. Brake lights and turn-signal malfunctions have been cited as having caused rear-end collisions.
◆ A small grass fire erupted at a construction site. When the operator of the vehicle nearest to the fire attempted to extinguish the blaze, he discovered that the fire extinguisher had already been discharged.

When the operator fails to report deficiencies to the mechanics, dangerous consequences may result.

The Solution
The goal of any maintenance program is to maintain the company equipment so that the daily tasks can be performed safely and on schedule. Since the operator is using the equipment on a regular basis, he or she is in the position to spot potential problems before they become serious. For a successful maintenance program, the following recommendations should be implemented:

◆ Hold a mandatory four-hour equipment maintenance training class conducted by mechanics in the motor pool. This training would consist of a hands-on approach to preventive maintenance checks and services at the operator level.
◆ Require operators to perform certain checks on a vehicle before checking it out of the motor pool. A vehicle checklist would be turned in to maintenance personnel.

The attached checklist would require 5 to 10 minutes to complete.

Conclusion
I believe the cost of maintaining the vehicle fleet at Apex will be reduced when potential problems are detected and corrected before they become serious. Operator training and the vehicle pretrip inspection checklist will ensure that preventable accidents are avoided. I will call you this week to answer any questions you may have about this proposal.

For practice with organization do the Patterns of Organization Activity.

Apex Transportation and Equipment
Fleet Maintenance Division
Vehicle Checklist
Pretrip Inspection

Inspected by: _____   Date: _____

Vehicle #: _____   Odometer: _____

Fluid Levels, Full/Low   Comments

_____ Engine Oil   _____

_____ Transmission Fluid   _____

_____ Brake Fluid   _____

_____ Power Steering   _____

_____ Radiator Level   _____

Before Cranking Vehicle

_____ Tire Condition   _____

_____ Battery Terminals   _____

_____ Fan Belts   _____

_____ Bumper and Hitch   _____

_____ Trailer Plug-in   _____

_____ Safety Chains   _____

After Cranking Vehicle

_____ Parking Brakes   _____

_____ Lights   _____

_____ All Gauges   _____

_____ Seat Belts   _____

_____ Mirrors/Windows/Wipers   _____

_____ Clutch   _____

_____ Fire Ext. Mounted and Charged   _____

_____ Two-Way Radio Working   _____

Additional Comments:_____

_____

❓ 9. Ethics Assignment

For this assignment your instructor will place you into a team, with the goal of presenting an oral or written report.

Using the Internet, find the Code of Ethics for the Association for Computing Machinery (ACM), the Institute of Electrical and Electronics Engineers (IEEE), the National Society for Professional Engineers (NSPE), the Society for Technical Communication (STC), or other professional organization. Evaluate the quality, usefulness, and appropriateness of the organization's guidelines by answering the following questions:

a. What do the guidelines suggest about the role of technical communicators in the workplace?

b. How would you adjust the depth, breadth, or balance of the items presented, if at all?

c. How does the document satisfy, or fail to satisfy, the ethical guidelines discussed in this chapter?

d. Are all guidelines and terms clear to the reader?

10. International Communication Assignment

World cultures differ in the way they organize information and in the visual cues they use for readers. Using a resource like www.newsdirectory.info or www.newspapers.com, find Web sites for newspapers from three different countries and analyze each Web site for the way information is presented. Do you notice any differences in how information is arranged on the pages of the site? For example, are the Web site's topics arranged vertically on the left of the page, as they are in most English-language Web sites? How are graphics treated? What other differences do you see? Write a brief essay about what you have learned about how these Web sites organize information, and what issues companies that are creating Web sites for global audiences need to be aware of.

Chapter | 2 | Collaboration and Writing

>>> Chapter Outline

In the workplace, correspondence and some short documents may be written by a single author, but most documents are the result of some kind of collaboration between writers. In one study, technical communication managers identified the two most important competencies for technical communicators as the ability to collaborate with subject matter experts and the ability to collaborate with coworkers.[1] This collaboration may be as short term as asking a coworker to read through a report before turning it in to a supervisor or a longer commitment as a member of a standing team that creates documents. You may collaborate with others in the development and delivery of products or services, in the marketing of those products or services, or in creating documentation to support those products or services. *Collaborative writing* (also called *team writing*) can be defined as follows:

> Collaborative writing: the creation of a document by two or more people. Documents are created collaboratively to meet the common purposes and goals of a community of writers, editors, and readers.

This chapter focuses on collaboration strategies as they are used in the writing process, but many of these strategies can contribute to the success of any team project.

>>> Approaches to Collaboration

The scope of the writing project, the setting in which it is written, and the number of people involved can all influence the form that collaborative writing takes. There are five common approaches to writing collaboratively:

■ **Divide and conquer:** When the writing project is large and has clearly defined sections, it may be helpful to assign individual sections of the document to specific writers. Later in the process the parts of the document are brought together and combined. Many documents today are produced using a version of this approach that depends on modular writing, discussed later in this chapter.

■ **Specialization:** Often referred to as writing in *cross-functional teams,* this version of divide and conquer assigns the parts of the project to team members based on their specialties. On a proposal writing team, for example, an engineer might write the technical descriptions and specifications, an accountant might write the budget projections, an account representative might write the more persuasive sections of the proposal, and a technical communicator might provide the overall design for the document and assemble the parts.

■ **Sequence:** In this approach, several people are involved in creating a document. Instead of working on it at the same time, however, they pass it from one person to the next. An engineer may write a description of a new product and then hand it off to an artist who creates images. The description and images then are sent to a documentation specialist, who revises the description for readers who do not have the engineer's expertise

[1]Kenneth T. Rainey, Roy K. Turner, and David Dayton, "Do Curricula Correspond to Managerial Expectations? Core Competencies for Technical Communicators," Technical Communication 52, no. 3 (2005): 323–352.

and integrates the text and images. Finally, the documentation specialist may pass along the document to a marketing communication writer, who uses it to create a description of the product for the company's Web site.

■ **Dialog:** When two writers are working together on a project, they may work best by sending drafts back and forth to each other, commenting and revising until they are both pleased with the final draft. This practice is common in settings where supervisors comment on the documents that their employees write, or when a writer is collaborating with an editor. When writing in this back-and-forth dialog, it is important to keep versions of each draft separate, in case the writers decide that an earlier version was more appropriate for the document's purpose.

■ **Synthesis:** This approach to team writing works best with two or three writers, and with shorter documents. The team writes together at the same time, adding ideas and commenting on the work as it progresses. They may work together at the same computer, or they may work on the same document simultaneously from different locations, using software that is designed for collaboration. This is the most seamlessly collaborative approach to writing, and it is most successful when the members of the team have worked together long enough to know each other well.

>>> Collaboration and the Writing Process

Writing collaboratively uses the same steps in the writing process as those used by individual writers. The team must identify the purpose of the document and the needs of its audience. It must collect information, plan the document, draft, and revise. And the team must do these tasks together, creating a cohesive and useful document.

The Writing Team

Some organizations have standing teams for common types of projects such as proposal writing, or for ongoing projects such as compliance with regulations. Teams also may be temporary, coming together for one project, and then separating, each member moving on to another project. Whether the team is a permanent (or standing) team or one that has been brought together for a single project, be aware of the individual roles that team members play. Begin by identifying the skills that each member can contribute to the project and assign tasks based on those skills. Do not just assume that skills are limited to the team members' job titles. Effective teams include the following roles:

■ **The team leader** is the central contact person for team members and also the contact for people who are not on the team. This person also may be working as the project manager.

■ **The planning coordinator** is responsible for managing communication among team members, for keeping track of benchmarks and deadlines, and for preparing for meetings. On small teams, the team leader may serve as the planning coordinator.

■ **The archivist** keeps minutes of meetings, copies of all written communication, and copies of all other written material related to the project. At the end of the project, the archivist creates the material that is stored in the organization's library or archives.

- **Devil's advocate** is a role that often occurs spontaneously, as one member of a team raises concerns or points out problems. She or he helps avoid *groupthink*, when members of a group begin to echo each other and stop looking critically at the work they are doing. Some teams formally assign this role, rotating it from meeting to meeting. If you find yourself raising concerns about a project during a meeting, it is helpful to announce it—"I'm just playing devil's advocate here, but …"—as a way of keeping the focus on the project and avoiding the temptation to make disagreements personal.

Planning

As with any writing project, team projects must be planned carefully. The Planning Form in the back of this book can be used for team writing in the same way that you can use it for individual writing projects.

Begin by identifying your audience. Who will be reading this document? What do they expect to learn from it? You should also identify the stakeholders in your team project. Obviously, the team members themselves have a stake in the success of the project, but others may be interested in its success, such as members of management, employees in other departments, and the organization as a whole. Clients are important stakeholders, especially if they have hired your organization for the project on which your team is working. If you are working on a client's project, you should work closely with the client and consider the client's representative a member of your team.

As part of the planning process, you must state clearly the desired outcome of the project. How will you know if you have completed it successfully? Your team's goal should be more than simply producing the required document. You should decide what information makes that document successful, where to find the information, and how best to organize the information. Then identify the tasks that must be accomplished to achieve the project's goals and assign the tasks to team members.

Budgeting Time and Money

Once you have identified tasks to be accomplished, you should identify *benchmarks*—the deadlines for specific tasks that keep the project focused and on schedule. These benchmarks vary from project to project, but common benchmarks for writing projects include the following:

- Completion of preliminary research
- Organization of collected information
- Planning of graphics
- Completion of first draft
- Editing of late draft
- Document design
- Publication of document

After identifying the benchmarks, your team can plan the calendar for the project. It is rare for a team to be able to set its own deadline. Team projects usually have a deadline that has been imposed from outside, so it is helpful to *backplan* the schedule for the project. Backplanning begins with the due date and works backwards. For example, if a project is due on July 1, the project coordinator may ask how long it will take to complete the final edit on the document. If it will take two days, then the benchmark to have the draft ready for final editing is two days before the due date. Working backward through the benchmarks that the team has identified, the project manager plans the rest of the schedule.

Using Schedule Charts

Schedule charts provide a graphic representation of a project plan. Many documents, especially proposals and feasibility studies, include a schedule chart to show readers when specific activities will be accomplished. Often called a milestone or Gantt chart (after Henry Laurence Gantt, 1861–1919), it usually includes these parts (Figure 2-1):

- **Vertical axis,** which lists the various parts of the project, in sequential order
- **Horizontal axis,** which registers the appropriate time units
- **Horizontal bar lines** (Gantt) or separate markers (milestone), which show the starting and end dates.

Communication

Face-to-face meetings are the best way to keep a team running smoothly. Today, however, many teams are spread across different company branches and even different countries, so this is not always possible. However, it is beneficial if teams can meet in person at least once at the beginning of a project and once near the end of a project.

Computers can be used to overcome many obstacles for writers and editors in different locations. Indeed, electronic communication can help accomplish all the guidelines in this chapter. Specifically, (1) e-mail can be used by group members to get to know each other; (2) e-mail or a computer conference can be used to establish goals and ground rules; (3) synchronous, or real-time, groupware can help a team brainstorm about approaches to the project (and may, in fact, encourage more openness than a face-to-face brainstorming session); (4) computer conferences combined with groupware can approximate the storyboard process; and (5) either synchronous or asynchronous groupware can be used to approximate the editing process.

■ **Figure 2-1** ■
Schedule chart variations

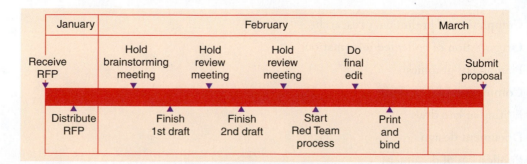

■ **Figure 2-1** ■
Continued

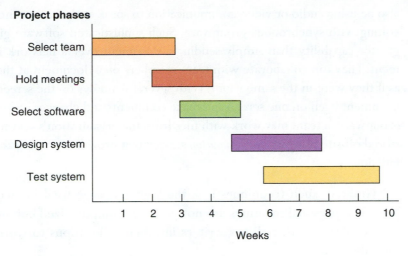

Project phases

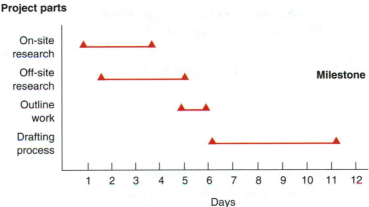

Project parts

■ **E-mail:** Team members can send and receive messages from their office computers or from remote locations. They can also attach documents in a variety of forms. When attaching a document to an e-mail, you should identify by file name and type of document (e.g., as a PDF) in the e-mail.

■ **Computer conference:** Members of a team can make their own comments and respond to others' comments on a specific topic or project. Computer conferences may be open to all interested users or open only to a particular group. For the purposes of collaborative writing, the conference probably would be open only to members of the writing team. A leader may be chosen to monitor the contributions and keep the discussion focused. Contributions may be made over a long period, as opposed to a conventional face-to-face meeting where all team members are present at the same time. Accumulated comments in the conference can be organized or indexed by topic. The conference may be used to brainstorm and thus to generate ideas for a project, or it may be used for comments at a later stage of the writing project.

■ **Groupware:**   Team members using this software can work at the same time, or at different times, on any part of a specific document. Groupware that permits contributions at the same time is called *synchronous*; that which permits contributions at different times is called *asynchronous*. Because team members are at different locations, they may also be using audio or video communication to speak at the same time they are writing or editing with synchronous groupware. Such sophisticated software gives writers a much greater capability than simply sending a document over a network for editing or comment. They can collaborate with team members on a document at the same time, almost as if they were in the same room. With several windows on the screen, they can view the document itself on one screen and make comments and changes on another screen. Using groupware, a team may work with files from the organization's server, or the files may be stored off site using a *cloud computing* service that provides server space for businesses and individuals.

Granted, such techniques lack the body language used in face-to-face meetings. Yet when personal meetings are not possible, computerized communication can provide a substitute that allows writers in different locations to work together to meet their deadline.

Of course, computers can create problems during a group writing project, if you are not careful. When different parts of a document have been written and stored by different writers, your group must be vigilant during the final editing and proofreading stages. Before submitting the document, review it for consistency and correctness.

Modular Writing

In the past, team members of a collaborative writing project could assume that before the final version of the document was released, they would have a chance to review the entire document. Today, however, the writing process in organizations is changing. Documents are broken into small sections, with different people responsible for each section. Variations of this practice go by many names—single sourcing, structured authoring, or content management. In this book, we refer to the general process as *modular writing*.

> **Modular writing:** A process in which large documents are broken down into smaller elements, and different people are given responsibility for each element. These smaller elements are usually stored electronically so that they can be retrieved and edited or assembled into larger documents, help files, or Web pages as they are needed.

For example, in a company that produces a number of documents for maintenance equipment, several people may share responsibility for all of the documents at once. One writer may be responsible for technical descriptions and another for instructions. An engineer may be responsible for technical specifications, and a graphic artist may be responsible for schematics and illustrations. Each person saves his or her work on a server where it can be accessed by anyone who uses it in a document. Someone writing a proposal to sell the equipment to a client may use the technical description and specifications. Each user's manual can be assembled from the elements that are specific to the equipment and to the user's needs. If the company sells its products overseas, translators in other countries can begin working on sections of a user's manual as soon as the individual sections are saved to the server, instead of waiting to receive the whole document before translation begins.

■ Figure 2-2 ■
Example of module
with conditional
text

About Us

M-Global, Inc., was founded in 1963 as McDuff, Inc., by Rob McDuff, as a firm that specialized in soils analysis. Since then, [we have][M-Global has] added hazardous waste management and clean-up, equipment development, business services, and documentation services.

[Our][M-Global's] teams [ensure][have ensured] compliance with construction codes and quality of materials in road, dam, and building construction projects such as the Nevada Gold Dome, with a savings to[our][the] client of $100,00. [We protect][M-Global has protected] threatened and endangered ecosystems by conducting rigorous environmental impact studies. [We help][M-Global's consultants have helped] organizations improve their internal operations and client services—in 2008, Kansans for Security and Privacy awarded [us][M-Global] the Peace of Mind Award for [our][M-Global's] work with the Kansas Department of Social and Health Services security protocols.

Today, after almost 50 years of business, M-Global, Inc., has about 2,500 employees. There are nine offices in the United States and six overseas, as well as a corporate headquarters in Baltimore. What started as a technical consulting engineering firm has expanded into a firm that does [quality][both] technical and non-technical work for a variety of customers.

Figure 2-2 is an example of modular writing. This introduction to the company includes conditional text that is coded for use in different types of documents. The information that appears in blue can be used in marketing materials like sales brochures or the company Web site. The information that appears in red is used in more formal documents like reports and proposals.

Modular writing requires careful planning. The writing team must identify all the modules needed in the final project and assign those modules to different writers. Individual writers may never see a draft of the complete document. To ensure consistency throughout all documents created from the separate elements, the writing team must first create a thorough style guide and adhere to it, even if the team includes an editor whose job is to check all documents for consistency.

Although organizations that use modular writing face many challenges, it has benefits that make the effort worthwhile. If a product is improved, the modules that are affected by the change can be updated easily. Then, any documentation about the product includes accurate information automatically. In the "About Us" text in Figure 2-2, new information about the organization's accomplishments can be added to the source module, after which all documents using this text are automatically updated before being printed or published. Because all the updates are kept in one file, there is no problem with someone missing an important update.

>>> Teamwork

Whether you are an engineer creating a document with other engineers, a technical communicator assigned to a company branch, or a documentation specialist on a cross-functional team, you should understand and stay focused on the project goals. This section offers six pointers for team writing to be used in this course and throughout your career. The suggestions concern the writing process as well as interpersonal communication.

Guidelines for Team Writing

>> Team Guideline 1: Get to Know Your Team

Most people are sensitive about strangers evaluating their writing. Before collaborating on a writing project, learn as much as you can about those with whom you will be working. Drop by their offices before your first meeting, or talk informally as a group before the writing process begins. In other words, establish a personal relationship. This familiarity helps set the stage for the spirited dialog, group criticism, and collaborative writing to follow.

>> Team Guideline 2: Set Clear Goals and Ground Rules

Every writing team needs a common understanding of its objectives and procedures for doing business. Either before or during the first meeting, the following questions should be answered:

1. What is the team's main objective?
2. Who will serve as team leader?
3. What exactly will be the leader's role in the group?
4. How will the team's activities be recorded?
5. How will responsibilities be distributed?
6. How will conflicts be resolved?
7. What will the schedule be?
8. What procedures will be followed for planning, drafting, and revising?

>> Team Guideline 3: Use Brainstorming Techniques for Planning

The term *brainstorming* means to pool ideas in a nonjudgmental fashion. In this early stage, participants should feel free to suggest ideas without criticism by colleagues in the group. This nonjudgmental approach does not come naturally to most people; thus, the leader may have to establish ground rules for brainstorming before the team proceeds.

Following is one sample approach to brainstorming:

Step 1:   The team recorder takes down ideas as quickly as possible.

Step 2:   Ideas are written on large pieces of paper affixed to walls around the meeting room so all participants can see how major ideas fit together.

Step 3: Members use the recorded ideas as springboards for suggesting other ideas.

Step 4: The team takes some time to digest ideas generated during the first session, before meeting again.

>> Team Guideline 4: Use Storyboarding Techniques for Drafting

Storyboarding helps propel participants from the brainstorming stage toward completion of a first draft. It also makes visuals an integral part of the document. Originating in the screenwriting trade in Hollywood, the storyboard process can take many forms, depending on the profession and individual organization. In its simplest form, a *storyboard* can be a sheet of paper or an electronic template that contains (1) one draft-quality illustration and (2) a series of sentences about one topic. As applied to technical writing, the technique involves six main steps:

Step 1: The team or its leader assembles a topic outline from ideas brought forth during the brainstorming session.

Step 2: All team members are given one or more topics to develop on storyboard forms.

Step 3: Each member works independently on the boards, creating an illustration and a series of subtopics for each main topic.

Step 4: Members meet again to review all completed storyboards, modifying them where necessary and agreeing on key sentences.

Step 5: Individual members develop draft text and related graphics from their own storyboards.

Step 6: The team leader or the entire group assembles the draft from the various storyboards.

>> Team Guideline 5: Agree on a Revision Process

As with drafting, all members usually help with revision. Team editing can be difficult, however, as members strive to reach consensus on matters of style. Following are some suggestions for keeping the editing process on track:

- Avoid making changes simply for the sake of individual preference.
- Search for areas of agreement among team members, rather than areas of disagreement.
- Make only those changes that can be supported by accepted rules of style, grammar, and use.
- Ask the team's best all-round stylist to do a final edit.

This review will help produce a uniform document, no matter how many people work on the draft.

>> Team Guideline 6: Use Computers to Communicate

When team members are at different locations, computer technology can be used to complete some or even the entire project. Team members must have personal computers and the software to connect their machines to a network, allowing members to send and

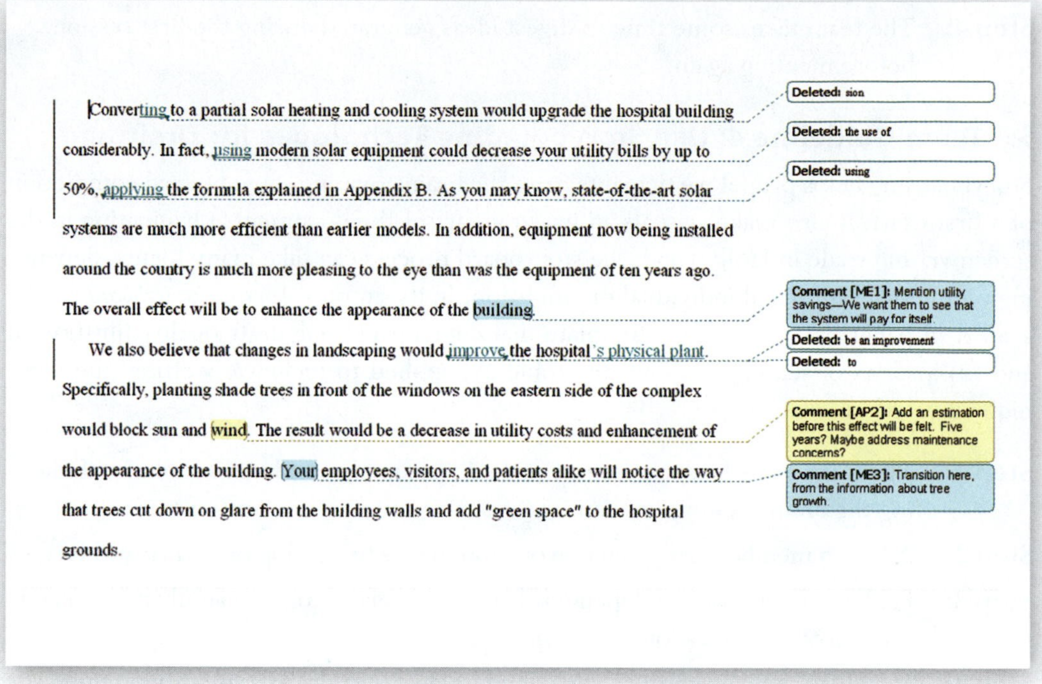

■ **Figure 2-3** ■
Edited text
showing team
members' markups
and comments

receive information online. Teams may also use common server space so that everyone has access to the most recent version of a project. Figure 2-3 shows how members of a proposal team have used the reviewing tools in their word processing program to make edits visible and raise questions or make suggestions about the draft. Notice that two different people have made comments, identified by two different sets of initials.

Collaboration will probably play an important part in your career. If you use the preceding techniques, you and your team members will build on each other's strengths to produce top-quality writing.

Writers and Subject Matter Experts

Many articles have been written about the importance of collaboration between technical communicators and the engineers, programmers, scientists, and other specialists with whom they work. These *subject matter experts* (*SMEs,* often pronounced "Smees") often contribute the technical content of documents, while technical communicators contribute their expertise in document design, writing, and editing. Good communication is important from the beginning of any project where technical communicators and SMEs are collaborating. The SME's misunderstanding of what technical communicators contribute to a project is one common cause of frustration for documentation specialists. However, the lack of technical knowledge on part of technical communicators can be source of frustration for SMEs. By keeping a few important guidelines in mind, technical communicators and SMEs can collaborate more effectively.

Guidelines for Collaborating with SMEs

>> Technical Communicator Guideline 1: Use the SME's Time Wisely

Do your background research before contacting the SME. Do not waste the specialist's time with questions that can be answered through other sources.

>> Technical Communicator Guideline 2: Put Questions in Writing When Possible

Make sure that e-mail questions are clear. You will not get useful answers if your questions are ambiguous or confusing.

>> Technical Communicator Guideline 3: Prepare for Interviews and Meetings

Have clear goals. If you want to ask for feedback on documentation, send it to the SME beforehand and bring a copy with you.

>> Technical Communicator Guideline 4: Treat the SME with Respect

When you are making changes in text that has been supplied by a technical specialist, remember that you are reading a draft, not a polished document. Never make negative comments to other employees (including fellow writers), about the writing ability of SMEs.

Guidelines for Being a Collaborative SME

>> SME Guideline 1: Keep Technical Communicators Informed

Provide technical communicators with the information they need, even if they do not ask for it. This includes keeping them informed of changes or updates of products or projects that they are documenting.

>> SME Guideline 2: Respond to E-mails and Phone Calls Promptly

If you are not sure what is being requested, ask for clarification. If you are being asked to a meeting or an interview, make time for it. Delays in providing necessary information to a documentation specialist can delay an entire project.

>> SME Guideline 3: Prepare for Interviews and Meetings

Find out ahead of time what you are going to be asked to explain or provide. Have all appropriate prototypes, samples, or products on hand, if possible. If something comes up that you cannot answer right away, make a note of it and respond as soon as possible.

>> SME Guideline 4: Treat the Technical Communicator with Respect

If the technical communicator has revised text that you provided, this is not a criticism of your writing ability. The changes were probably made to shift the focus of the text to the users' needs. Clear documentation is an important part of a well-run organization, as well as being important to the products or services that your organization provides to its clients.

>>> Chapter Summary

Most organizations rely on people collaborating throughout the writing process to produce documents. The success of writing projects depends on information and skills contributed by varied employees. Collaborative writing can be done by pairs of people working side by side or by several people each contributing their own expertise. Writing teams can have standing tasks, such as proposal writing, or they can be put together for a single project. In some organizations, most print and electronic documents, even Web sites, are created collaboratively through modular writing.

In collaborative writing, the whole is greater than the sum of the parts. In other words, benefits go beyond the collective specialties and experience of individual group members. Participants create new knowledge as they plan, draft, and edit their work together. They become better contributors and faster learners simply by being a part of the social process of a team. Discussion with fellow participants moves them toward new ways of thinking and inspires them to contribute their best. This collaborative effort yields ideas, writing strategies, and editorial decisions that result from the mixing of many perspectives.

Of course, team writing does have drawbacks. Most notably, the group must make decisions without falling into time traps that slow down the process. There must be procedures for getting everyone's ideas on the table and for reaching decisions on time. A leader with good interpersonal skills helps the group reach its potential, whereas an indecisive or autocratic leader is an obstacle to progress. Good leadership rests at the core of every effective writing team.

In addition to good leadership, shared decision-making is at the heart of every successful writing team. Members of the writing team must communicate information and expectations clearly. Participants in a team must work together during the planning, drafting, and revising stages of writing. They must respect each other and remember that successful writing projects contribute to the success of their organization.

>>> Learning Portfolio

Collaboration at Work Advice about Advising

General Instructions

Each Collaboration at Work exercise applies strategies for working in teams to chapter topics. The exercise assumes you (1) have been divided into teams of about three to six students, (2) will use time inside or outside of class to complete the case, and (3) will produce an oral or written response.

Background for Assignment

Academic advising can be one of the most important as well as the most confusing activities for college students, especially for students who are going through the process for the first time. Students depend on advice from other students, from seminars and workshops, and from teachers. This advice may not always fit the student's situation, or steps for advising and enrollment may change. This assignment asks you to create a document or Web page to help your fellow students get the most out of advising.

Team Assignment

In your groups, brainstorm the questions that you have had about advising and enrollment. What advice would

you give to fellow students? Identify the steps in the process, where information is currently available, and other sources of information (such as faculty members, the Registrar's Office, or Student Services). Your instructor may assign a team leader, or ask each team to choose its own leader. Decide what information you must gather, how you will gather it, and how you are going to make it available to students. Your team should also decide what approach it will take to gathering and writing the information—divide and conquer, writing in sequence, or working at the same time (see p. 42).

Your instructor may decide to make this an assignment in modular writing. If so, the class can brainstorm about the content and sources of information, and then teams will be assigned specific tasks. One team will be responsible for creating style guidelines and a document template to ensure a uniform voice and appearance throughout the document. This team, or another team (depending on how large the class is), will also have responsibility for final editing on the project. Other teams will be assigned to gather information from various sources and to write specific sections of the document. Your instructor will help you decide how your project will be made available to students.

Assignments

1. Survey—Your Experience with Teams

Answer the following questions about your experience collaborating on projects, either in school or at work. In teams of five or more, compile and present information in a meaningful way. Discuss the responses.

A. Briefly describe your experiences with the following:

- Divide and conquer: The team planned the project together and randomly assigned tasks to each member.
- Specialization: The team planned the project together and assigned tasks according to each person's expertise.
- Sequence: One person drafted the project, passed it along to the next person who revised the project, who passed it along to the next person, and so forth.
- Dialog: Two people worked on the project; one drafted it and gave it to the other, who revised it and returned it

for more revisions, until both partners were happy with the result (or until the deadline).
- Synthesis: Two or three people created the project together, working side by side. Every responsibility in the project was shared completely.

B. What makes a good member of a project team?

C. What problems have you encountered in collaborative projects?

2. Schedule Charts

Create a schedule chart that reflects your work on one of the following:

- A project at work
- A laboratory course at school
- A lengthy project such as the one for which you are using this book

All of the following assignments should be completed in teams of four or five students.

3. Short Report

In teams, write a brief evaluation of the teaching effectiveness of either the room in which your class is held or some other room or building of your instructor's choice. In following the tasks listed in this chapter, the team must establish criteria for evaluation, apply these criteria, and report on the results.

Your brief report should have three parts: (1) a one-paragraph summary of the room's effectiveness, (2) a list of the criteria used for evaluation, and (3) details of how the room met or did not meet the criteria you established.

Besides preparing the written report, be prepared to discuss the relative effectiveness with which the team followed this chapter's guidelines for collaborative writing. What problems were encountered? How did you overcome them? How would you do things differently next time?

4. Computerized Communication

If your campus computer facilities permit, set up a groupware folder with members of a writing team to which you have been assigned by your instructor. Decide on a topic on which you and your team members will write. Each team member should post one short document to the folder, and each team member should contribute to the other documents in the folder. Print the contents of the team's folder and submit it to your instructor. Depending on the instructions you have been given, this assignment may be independent or it may be related to a larger collaborative writing assignment.

5. Research and Presentation

Using the working groups your instructor has established, collect information on *collaborative learning* and then make a brief oral presentation on your findings to the entire class. Your sources may involve print media or computer sources, such as the Internet.

6. Sequential Collaboration

Working in pairs, create a presentation about writing in teams. Use the information from this chapter, as well as information from other courses or from Internet sources. One student should write the presentation notes, and then give them to the other student who will create slides in presentation software (like PowerPoint) to accompany the notes. When you are done, practice through the presentation. How well do the slides work with the text of the notes? What problems did you encounter in the project? Turn in your

presentation and a paragraph discussing this collaboration experience.

7. Cloud Computing

Using the working groups your instructor has established, collect information on business etiquette or on another topic that your instructor assigns. As a team, create a set of business etiquette guidelines. Use a free cloud computing tool, such as GoogleDocs (https:docs.google.com) or Drop-Box (www.dropbox.com) to write your guidelines. Include an Abstract that introduces the guidelines, a Body that lists the guidelines, and a Conclusion that explains why the guidelines are important. Share the online document with your instructor.

8. Reviewing Tools

For this assignment, your instructor will assign pairs or small teams. Write your responses to the questions in Assignment 1 in paragraph form. Send your draft of this document as an e-mail attachment to your partner or team members. When you have received the document from your partner or team members, suggest changes, ask questions, and add comments using the track changes and commenting tools in your word processing program. Save the document with a new name and send it to the writer as an e-mail attachment. Your instructor may ask you to comment on your experience with the reviewing tools.

9. Ethics and Collaboration

Create an evaluation sheet that could be used for any collaborative projects that your instructor assigns during the semester. Decide if the whole group should sign one document or if individuals should write their own. Explain your decision in a cover memo to your instructor.

10. International Communication Assignment

This chapter offers guidelines on team writing because collaborative communication is essential for success in most careers. However, world cultures differ in the degree to which they use and require collaboration on the job. For this assignment, interview someone who is from a culture different from your own. Using information supplied by this informant, write a brief essay in which you (1) describe the importance of collaboration in the individual's home culture and workplace, (2) give specific examples of how and when collaborative strategies would be used, and (3) modify or expand this chapter's "Guidelines for Group Writing" to suit the culture you are describing, on the basis of suggestions provided by the person you interviewed.

Chapter | 3 | Visual Design

>>> Chapter Outline

As with the organizing principles discussed in Chapter 1, good visual design can help your readers find the information that they need. This chapter covers the visual elements of technical documents, another basic building block in technical communication. An operating definition is as follows:

Visual design: A term that refers to creating clear, readable, and visually interesting documents through judicious use of page design, including white space, headings, lists, font choice, and of graphic design elements.

This chapter presents guidelines and examples for visual design, as well as commentary on the use of computers in the design process.

>>> Elements of Page Design

In the workplace, readers are busy, and few take the time to read a document from cover to cover. Some documents, such as manuals, are used as reference works, only consulted as a last resort. Your challenge is to make your documents inviting by making pages interesting to the eye. A document may offer your reader great opportunities but never get read. Why? Because it does not look inviting.

Good organization, as pointed out in Chapter 1, can fight readers' indifference by giving information when and where they want it. However, to get and keep readers interested, you must use effective page design—on each page of your document. Each page needs the right combination of visual elements to match the needs of your readers and the purpose of the document.

Good page design can work with good organization to help readers find information that they need in a document. Readers can recognize important information by its location on a page, by use of contrast, or by repetition of identically formatted elements, such as guidelines or warning boxes.

Many firms know the benefits of visual imaging so they develop company style sheets for frequently used documents, including letters, memos, various types of reports, and proposals. Once developed, these style sheets are assembled into a style manual and distributed for general use. To make universal use easier, they are often loaded as templates or styles into the company's text editing software. If the company you work for does not use a style manual, use the elements of page design shown in this chapter to develop your own style sheets. More information about templates and style sheets appears on pages 66 to 69 at the end of this chapter.

White Space

The term *white space* simply means the open places on the page with no text or graphics—literally, the white space (assuming you are using white paper). Experts have learned that readers are attracted to text because of the white space that surrounds it, as with a newspaper advertisement that includes a few lines of copy in the middle of a white page. Readers connect white space with important information.

In technical communication you should use white space in a way that (1) attracts attention, (2) guides the eye to important information on the page, (3) relieves the boredom of reading text, and (4) helps readers organize information. Here are some opportunities for using white space effectively:

1. **Margins:** Most readers appreciate generous use of white space around the edges of text. Marginal space tends to frame your document, so the text does not appear to push the boundaries of the page. Good practice is to use 1- to 11/2-inch margins, with more space on the bottom margin. When the document is bound, the margin on the edge that is bound should be larger than the outside margin to account for the space taken by the binding (Figure 3-1.)

2. **Hanging Indents:** Some writers place headers and subheads at the left margin and indent the text block an additional inch or so, as shown in Figure 3-2. Headers and subheads force the readers' eyes and attention to the text block. Another common use of hanging indents is bulleted and numbered lists.

3. **Line Spacing:** When choosing single, double, or 11/2 line spacing, consider the document's degree of formality. Letters, memos, short reports, and other documents read in one sitting are usually single-spaced. Longer documents, especially if they are formal, are usually 11/2-space or double-spaced, sometimes with extra spacing between paragraphs. Manuscripts or documents that will be typeset professionally are usually double-spaced.

■ **Figure 3-1** ■ Use of white space: margins

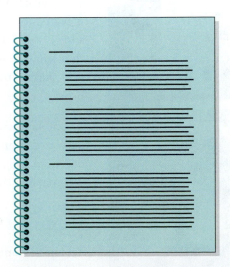

■ **Figure 3-2** ■ Use of white space: hanging indents

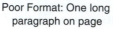

Poor Format: One long
paragraph on page

Better Format: Several
paragraphs on page

4. **Paragraph Length:** New paragraphs give readers a chance to regroup as one topic ends and another begins. These shifts also have a visual impact. The amount of white space produced by paragraph lengths can shape reader expectations. For example, two long paragraphs suggest a heavier reading burden than do three or four paragraphs of differing lengths. Thus, it is helpful to break complex information into shorter paragraphs. Many readers skip long paragraphs, so vary paragraph lengths and avoid putting more than 10 lines in any one paragraph (Figure 3-3).

5. **Heading Space and Ruling:** White space helps the reader connect related information immediately. Always have slightly more space above a heading than below it. That extra space visually connects the heading with the material into which it leads. In a double-spaced document, for example, you would add a third line of space between the heading and the text that came before it. In addition, some writers add a horizontal line across the page above headings, to emphasize the visual break.

In summary, well-used white space can add to the persuasive power of your text. As with any design element, however, it can be overused and abused. Make sure there is a reason for every decision you make with regard to white space on your pages.

Lists

Technical communication benefits from the use of lists. Readers welcome your efforts to cluster items into lists for easy reading. In fact, almost any group of three or more related points can be made into a bulleted or numbered listing. Following are some points to consider as you apply this important feature of page design:

1. **Typical uses:** Lists emphasize important points and provide a welcome change in format. Because they attract more attention than text surrounding them, they are usually reserved for these uses:

> Examples
>
> Reasons for a decision
>
> Conclusions
>
> Recommendations
>
> Steps in a process
>
> Cautions or warnings about a product
>
> Limitations or restrictions on conclusions

2. **Number of items:** The best lists subscribe to the often-quoted rule that people retain no more than five to nine items in their short-term memory. A listing of more than nine items may confuse rather than clarify an issue. Consider placing 10 or more items in two or three groupings, or grouped lists, as you would in an outline. This format gives the reader a way to grasp information being presented.

3. **Use of bullets and numbers:** The most common visual clues for listings are numbers and *bullets* (enlarged dots or squares like those used in the following listing). Following are a few pointers for choosing one or the other:

- **Bullets.** Best in lists of five or fewer items, unless there is a special reason for using numbers.

- **Numbers.** Best in lists of more than five items or when needed to indicate an ordering of steps, procedures, or ranked alternatives. Remember that your readers sometimes infer sequence or ranking in a numbered list.

4. **Format on page:** Every listing should be easy to read and pleasing to the eye. The following specific guidelines cover practices preferred by most readers:

- **Indent the listing.** Although there is no standard list format, readers prefer lists that are indented farther than the standard left margin. Five spaces are adequate.

- **Hang your numbers and bullets.** Visual appeal is enhanced by placing numbers or bullets to the left of the margin used for the list, as done with the items here.

- **Use line spaces for easier reading.** When one or more listed items contain over a line of text, an extra line space between listed items can enhance readability.

- **Keep items as short as possible.** Depending on purpose and substance, lists can consist of words, phrases, or sentences—such as the list you are reading.

Whichever format you choose, pare down the wording as much as possible to retain the impact of the list format.

5. **Parallelism and lead-ins:** Make the list easy to read by keeping all points grammatically parallel and by including a smooth transition from the lead-in to the listing itself. (The term *lead-in* refers to the sentence or fragment preceding the listing.) *Parallel* means that each point in the list is in the same grammatical form, whether a complete sentence, verb phrase, or noun phrase. If you change form in the midst of a listing, you take the chance of upsetting the flow of information.

Example: To complete this project, we plan to do the following:

- Survey the site
- Take samples from the three boring locations
- Test selected samples in our lab
- Report the results of the study

The listed items are in verb form (note the introductory words *survey, take, test,* and *report*).

6. **Punctuation and capitalization:** Although there are acceptable variations on the punctuation of lists, preferred usage includes a colon before a listing, no punctuation after any of the items, and capitalization of the first letter of the first word of each item. Refer to the alphabetized Handbook under "Punctuation: Lists" for alternative ways to punctuate lists.

In-Text Emphasis

Sometimes you want to emphasize an important word or phrase within a sentence. Computers give you these options: underlining, boldface, italics, and caps. The least effective are FULL CAPS and <u>underlining</u>. Both are difficult to read within a paragraph and distracting to the eye. The most effective highlighting techniques are *italics* and **boldface**; they add emphasis without distracting the reader.

Whatever typographical techniques you select, use them sparingly. They can create a busy page that leaves the reader confused about what to read. Excessive in-text emphasis also detracts from the impact of headings and subheadings, which should be receiving significant attention.

mytechcommlab

The Model Sales Proposal shows how elements of page design can call attention to important information.

>>> Elements for Navigation

As noted in Chapter 1, your audience will be busy and rarely read a longer report or document from the first page to the last. One way that you can help readers can locate the information they need is by using visual design to create navigational tools in your documents. You may be used to thinking of navigational tools in electronic texts such as Web pages or even PDFs, but print documents also use navigational devices such as tables of contents, indexes, headings, running headers and footers, and even color coding.

Headings

Headings are brief labels used to introduce each new section or subsection of text. They serve as (1) a signpost for the reader who wants to know the content, (2) a grabber to entice readers to read documents, and (3) a visual oasis of white space where the reader gets relief from text. Following are some general guidelines:

1. **Use your outline to create headings and subheadings.** A well-organized outline lists major and minor topics. With little or no change in wording, topics can be converted to headings and subheadings within the document.

2. **Use substantive wording.** Headings give readers an overview of the content that follows. They can determine whether readers read or skip the text. Strive to use concrete rather than abstract nouns, even if the heading must be a bit longer.

> *Original:* "Background"
>
> *Revised:* "How the Simmons Road Project Got Started" or "Background on Simmons Road Project"
>
> *Original:* "Costs"
>
> *Revised:* "Production Costs of the FastCopy 800" or "Producing the FastCopy 800: How Much?"

3. **Maintain parallel form in wording.** Headings of equal value and degree should have parallel grammatical form:

  A. *Headings That Lack Parallel Form*

      Scope of Services

      How Will Fieldwork Be Scheduled?

      Establish Contract Conditions

  B. *Revised Headings with Parallel Form*

      Scope of Services

      Schedule for Fieldwork

      Conditions of Contract

4. **Establish clear hierarchy in headings.** Whatever typographical techniques you choose, your readers must be able to distinguish one heading level from another.

■ **Consider using decimal headings for long documents.** Decimal headings include a hierarchy of numbers for every heading and subheading listed in the table of contents. Many an argument has been waged over their use. People who like them say that they help readers find their way through documents and refer to subsections in later discussions. People who dislike them say that they are cumbersome and give the appearance of bureaucratic writing.

Unless decimal headings are expected by your reader, use them only with formal documents that are fairly long. Following is the normal progression of numbering in decimal headings for a three-level document:

1.0 xxxxxxxxxxxxxx
  1.1 xxxxxxxxxx
    1.1.1 xxxxxxxxxx
    1.1.2 xxxxxxxxxx
  1.2 xxxxxxxxxx
    1.2.1 xxxxxxxxxx
    1.2.2 xxxxxxxxxx
2.0 xxxxxxxxxxxxxx
  2.1 xxxxxxxxxx
    2.1.1 xxxxxxxxxx
    2.1.2 xxxxxxxxxx
  2.2 xxxxxxxxxx
3.0 xxxxxxxxxxxxxx

Headers and Footers

As noted earlier, running headers and footers are important navigation devices in documents, and most word processing programs make the creation of headers and footers easy. In addition to being able to insert automatic page numbering, you can insert other information, such as short titles, your name, or your organization's name, on each page. You can decide where to position that information and you can hide it on selected pages. Some organizations put information, such as the computer filename or project identification number in document footers.

Color

Use of color, like fonts, focuses your reader's attention on important details. When used indiscriminately—inserted into a document just to show that color can be used—it is distracting.

Limit your use of color in routine documents for two reasons: (1) When you use professional printers, printing in color can be very expensive and (2) when you are using desktop printers, printing color documents can be very slow.

>>> Fonts

Your choice of size and type of font can affect how easy it is for your readers to find information, as well as how they interpret your tone and professionalism.

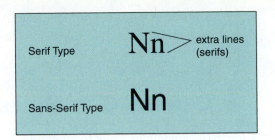

Font Types

Your choice of fonts may be either prescribed by your employer or determined by you on the basis of (1) the purpose of the document, (2) the image you want to convey, and (3) your knowledge of the audience.

Font types are classified into two main groups:

■ **Serif fonts:** Characters have "tails" at the ends of the letterlines.

■ **Sans-serif fonts:** Characters do not have tails (Figure 3-4).

If you are able to choose your font, the obvious advice is to use the one that you know is preferred by your readers. If you have no reader-specific guidelines, following are four general rules:

1. **Use serif fonts for regular text in your documents.** The tails on letters make letters and entire words more visually interesting to the reader's eye, and they reduce eye fatigue.

2. **Consider using another typeface—sans serif—for headings.** Headings benefit from a clean look that emphasizes the white space around letters. Sans-serif type helps attract attention to these elements of organization within your text.

3. **Avoid too many font variations in the same document.** Your rule of thumb might be to use no more than two fonts per document: one for text and another for headings and subheadings.

4. **Avoid unusual on novelty fonts.** To avoid problems in documents that are transmitted electronically, use fonts that are available on most computers. You should also avoid novelty fonts like Comic Sans or Broadway; these do not communicate a professional image.

Type Size

Most technical writing is printed in 10- or 12-point type. When you are choosing type size, however, be aware that the actual size of letters varies among font types. Some 12-point type appears larger than other 12-point type. Differences stem from the fact that your selection of a font affects (1) the thickness of the letters, (2) the size of lowercase letters, and (3) the length and style of the parts of letters that extend above and below the line. Before selecting your type size, run samples on your printer so that you are certain of how your copy will appear in final form.

With basics of page design foundation, you can decide to include graphics to clarify your document. The next sections provide general and specific guidelines for various graphics.

>>> General Guidelines for Graphics

mytechcommlab

Model Report 12: Final Report uses graphics effectively.

A few basic guidelines apply to all graphics. Keep the following fundamentals in mind as you move from one type of illustration to another.

>> Graphics Guideline 1: Determine the Purpose of the Graphic

Graphics, like text, should only be used if they serve a purpose. Ask yourself the following questions:

- What kind of visual information does your audience need to better understand the object, process, problem, or solution?
- What type of graphic can be used to present the data in the most interesting and informative way?
- Are any special symbols, colors, or font styles needed to reinforce the data?

>> Graphics Guideline 2: Evaluate the Accuracy and Validity of the Data

Unless the information you plan to include in your document is accurate, you run the risk of presenting information that could damage your credibility as well as the credibility of the document.

- Check the accuracy of information
- Make sure the source is reputable
- Ensure that data are not distorted by flawed scales or images

>> Graphics Guideline 3: Refer to All Graphics in the Text

With a few exceptions—such as cover illustrations used to grab attention—graphics should be accompanied by clear references within your text.

- Include the graphic number in Arabic, not Roman, when you are using more than one graphic
- Include the title, and sometimes the page number, if either is needed for clarity or emphasis
- Incorporate the reference smoothly into text wording
- Highlight significant information being communicated by the graphic

Following are two ways to phrase and position a graphics reference. In Example 1, there is the additional emphasis of the graphic's title, whereas in Example 2, the title is left out. Also, note that you can draw more attention to the graphic by placing the reference at the start of the sentence in a separate clause, or you can relegate the reference to a parenthetical expression at the end or middle of the passage. Choose the option that best suits your purposes.

- **Example 1:** In the past five years, 56 businesses in the county have started in-house recycling programs. The result has been a dramatic shift in the amount of property the county has bought for new waste sites, as shown in Figure 5.

- **Example 2:** As shown in Figure 5, the county has purchased much less land for landfills during the past five years. This dramatic reduction results from the fact that 56 businesses have started in-house recycling programs.

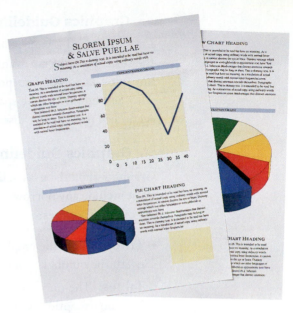

>> Graphics Guideline 4: Think About Where to Put Graphics

In most cases, locate a graphic close to the text in which it is mentioned. This immediate reinforcement of text by an illustration gives graphics their greatest strength. Variations of this option, as well as several other possibilities, include the following:

- **Same page as text reference:** A simple visual, such as an informal table, should go on the same page as the text reference if you think it too small for a separate page.

- **Page opposite text reference:** A complex graphic, such as a long table, that accompanies a specific page of text can go on the page opposite the text—that is, on the opposite page of a two-page spread. Usually this option is exercised *only* in documents that are printed on both sides of the paper throughout.

- **Page following first text reference:** Most text graphics appear on the page after the first reference. If the graphic is referred to throughout the text, it can be repeated at later points. (*Note:* Readers prefer to have graphics positioned exactly where they need them, rather than their having to refer to another part of the document.)

- **Attachments or appendices:** Graphics can go at the end of the document in two cases: first, if the text contains so many references to the graphic that placement in a central location, such as an appendix, would make it more accessible; and second, if the graphic contains less important supporting material that would only interrupt the text.

>> Graphics Guideline 5: Position Graphics Vertically When Possible

Readers prefer graphics they can view without having to turn the document sideways. However, if the table or figure cannot fit vertically on a standard $8\frac{1}{2} \times 11$-inch page, either use a foldout or place the graphic horizontally on the page. In the latter case, position the illustration so that the top is on the left margin. (In other words, the page must be turned clockwise to be viewed.)

>> Graphics Guideline 6: Avoid Clutter

Let simplicity be your guide. Readers go to graphics for relief from or reinforcement of the text. They do not want to be bombarded by visual clutter. Omit information that is not relevant to your purpose while still making the illustration clear and self-contained. Also, use enough white space so that the readers' eyes are drawn to the graphic. The final section of this chapter discusses graphics clutter in more detail.

>> Graphics Guideline 7: Provide Titles, Notes, Keys, and Source Data

Graphics should be as self-contained and self-explanatory as possible. Moreover, they must include any borrowed information. Follow these basic rules for format and acknowledgment of sources:

- **Title:** Follow the graphic number with a short, precise title—either on the line beneath the number or on the same line after a colon (e.g., "Figure 3: Salary Scales").
- **Tables:** The number and title go at the top. (As noted in Table Guideline 1 on page 55, one exception is informal tables, which have no table number or title.)
- **Figures:** The number and title usually go below the illustration. Center titles or place them flush with the left margin.
- **Notes for explanation:** When introductory information for the graphic is needed, place a note directly underneath the title or at the bottom of the graphic.
- **Keys or legends for simplicity:** If a graphic needs many labels, consider using a legend or key, which lists the labels and corresponding symbols on the graphic. For example, a pie chart might have the letters *A, B, C, D,* and *E* printed on the pie pieces and a legend at the top, bottom, or side of the figure listing what the letters represent.
- **Source information at the bottom:** You have an ethical, and sometimes legal, obligation to cite the person, organization, or publication from which you borrowed information for the figure. Either (1) precede the description with the word *Source* and a colon or (2) if you borrowed just part of a graphic, introduce the citation with *Adapted from.*

As well as citing the source, it is sometimes necessary to request permission to use copyrighted or proprietary information, depending on how you use it and how much you are using. (A prominent exception is most information provided by the federal government; most government publications are not copyrighted.) Consult a reference librarian for details about seeking permission.

>>> Specific Guidelines for Six Graphics

Illustrations come in many forms. Almost any nontextual part of your document can be placed under the umbrella term *graphic.* Among the many types, the following are often used in technical communication: (1) tables, (2) pie charts, (3) bar charts, (4) line charts, (5) flowcharts, and (6) technical drawings. This section of the chapter highlights their different purposes and gives guidelines for using each type.

Tables

Tables present readers with raw data, usually in the form of numbers but sometimes in the form of words. Tables are classified as either *informal* or *formal*:

- **Informal tables:** Limited data arranged in the form of either rows or columns
- **Formal tables:** Data arranged in a grid, always with both horizontal rows and vertical columns

The following five guidelines help you design and position tables within the text of your documents:

>> Table Guideline 1: Use Informal Tables as Extensions of Text

Informal tables are usually merged with the text on a page, rather than isolated on a separate page or attachment. As Figure 3-5 shows, an informal table usually has (1) no table number or title and (2) few if any headings for rows or columns. Also, it is not included in the list of illustrations in a formal document.

FTC staff then posted sets of three of these newly-created email addresses – consisting of an Unfiltered Address, an address at Filtered ISP 1, and an address at Filtered ISP 2 – on 50 Internet locations. The 50 Internet locations included websites controlled by the FTC[5] and several popular message boards, blogs, chat rooms, and USENET groups which had high hit/visit rates, according to ranking websites such as www.message-boards.com and Google popularity searches.[6] All of the 150 addresses were posted during a three day period in July 2005.

Graphic 1

Locations On Which E-mail Addresses
Were Posted

Type	Number
FTC Web site Pages	12
Message Boards	12
Blogs	12
Chat Rooms	12
USENET Groups	2

■ **Figure 3-5** ■
Informal table in a report
Source: Federal Trade Commission, *Email Address Harvesting and the Effectiveness of Anti-Spam Filters: A Report by the Federal Trade Commission's Division of Marketing Practices* (July 2010): 2.

>> Table Guideline 2: Use Formal Tables for Complex Data Separated from Text

Formal tables may appear on the page of text that includes the table reference, on the page following the first text reference, or in an attachment or appendix. In every case, you should do the following:

1. Extract important data from the table and highlight them in the text
2. Make every formal table as clear and visually appealing as possible by doing the following:
   - Use color to designate positive or negative totals, increases, or decreases or very important points.
   - Use gray screens (no denser than 10%–25%) to subordinate less-important data that appear on the table.

>> Table Guideline 3: Use Plenty of White Space

Used around and within tables, white space guides the eye through a table much better than black lines.

>> Table Guideline 4: Follow Usual Conventions for Dividing and Explaining Data

Figure 3-6 shows a typical formal table. It satisfies the overriding goal of being clear and self-contained. To achieve that objective in your tables, use the following guidelines:

1. **Titles and numberings:** Give a title to each formal table, and place the title and number above the table. Number each table if the document contains two or more tables.
2. **Headings:** Create short, clear headings for all columns and rows.
3. **Abbreviations:** Include in the headings any necessary abbreviations or symbols, such as *lb* or %. Spell out abbreviations and define terms in a key or footnote if the reader may need such assistance.

■ **Figure 3-6** ■
Example of a
formal table

TABLE 22: Employee Retirement Fund

Investment Type	Book Value	Market Value	% of Total Market Value
Temporary Securities	$ 434,084	434,084	5.9%
Bonds	3,679,081	3,842,056	52.4
Common Stocks	2,508,146	3,039,350	41.4
Mortgages	18,063	18,063	0.3
Real Estate	1,939	1,939	nil
Totals	$6,641,313	$7,335,492	100.0%

Note: This table contrasts the book value versus the market value of the Employee Retirement Fund, as of December 31, 2010.

4. **Numbers:** For ease of reading, round off numbers when possible. Also, align multi-digit numbers on the right edge, or at the decimal when shown.

5. **Notes:** Place any necessary explanatory headnotes either between the title and the table (if the notes are short) or at the bottom of the table.

6. **Footnotes:** Place any necessary footnotes below the table.

7. **Sources:** Place any necessary source references beneath the footnotes.

8. **Caps:** Use uppercase and lowercase letters rather than all caps.

Pie Charts

Familiar to most readers, *pie charts* show approximate relationships between the parts and the whole. Their simple circles with clear labels can provide comforting simplicity within even the most complicated report. Yet, the simple form keeps them from being useful when you must reveal detailed information or changes over time. Following are specific guidelines for constructing pie charts:

>> Pie Chart Guideline 1: Use Pie Charts Especially for Percentages and Money

Pie charts catch the readers' eyes best when they represent items divisible by 100—as with percentages and dollars. As Figure 3-7 shows, using pie charts for money breakdowns is made even more appropriate by the coinlike shape of the chart. *In every case, make sure your percentages or cents add up to 100.*

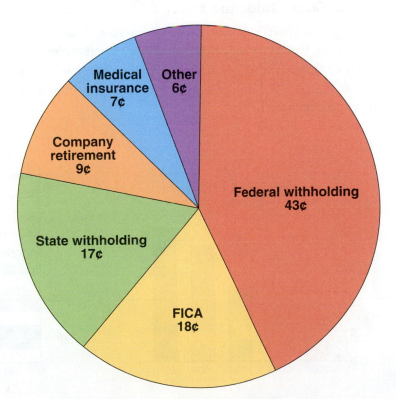

■ **Figure 3-7** ■ Pie chart showing money breakdown for average deductions from employees' paychecks

>> **Pie Chart Guideline 2:** Use No More Than Six or Seven Divisions

To make pie charts work well, limit the number of pie pieces to no more than six or seven. This approach lets the reader grasp major relationships without having to wade through the clutter of tiny divisions that are difficult to label and read.

>> **Pie Chart Guideline 3:** Move Clockwise from 12:00, from Largest to Smallest Wedge

Readers prefer pie charts oriented like a clock, with the first wedge starting at 12:00. Move from the largest to the smallest wedge to provide a convenient organizing principle, as in Figure 3-7.

>> **Pie Chart Guideline 4:** Draw and Label Carefully

The most common pie chart errors are (1) wedge sizes that do not correspond correctly to percentages or money amounts and (2) pie sizes that are too small to accommodate the information placed in them.

Bar Charts

Like pie charts, bar charts are easily recognized, because they are seen everyday in newspapers and magazines. Unlike pie charts, however, bar charts can accommodate a good deal of technical detail. Comparisons are provided by means of two or more bars running horizontally or vertically on the page. Use the following five guidelines to create effective bar charts:

>> **Bar Chart Guideline 1:** Use a Limited Number of Bars

Although bar charts can show more information than pie charts, both illustrations have their limits. Bar charts begin to break down when there are so many bars that information is not easily grasped. The maximum bar number can vary according to chart size, of course. Figure 3-8 shows two multi-bar charts. The impact of the charts is enhanced by the limited number of bars.

■ **Figure 3-8** ■ Bar charts

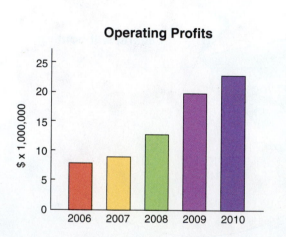

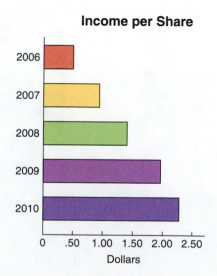

>> Bar Chart Guideline 2: Show Comparisons Clearly

Bar lengths should be varied enough to show comparisons quickly and clearly. Avoid using bars that are too close in length, because then readers must study the chart before understanding it. Such a chart lacks immediate visual impact.

>> Bar Chart Guideline 3: Keep Bar Widths Equal and Adjust Space between Bars Carefully

Although bar length varies, bar width must remain constant. As for distance between the bars, following are four options (along with examples in Figure 3-9):

- **Option A: Use no space** when there are close comparisons or many bars, so that differences are easier to grasp.
- **Option B: Use equal space, but less than bar width** when bar height differences are great enough to be seen in spite of the distance between bars.
- **Option C: Group related bars** to emphasize related data.
- **Option D: Use variable space** when gaps between some bars are needed to reflect gaps in the data.

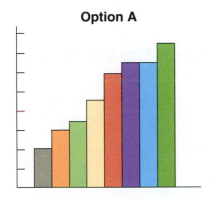

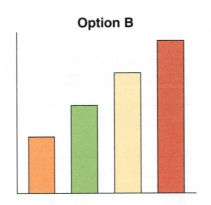

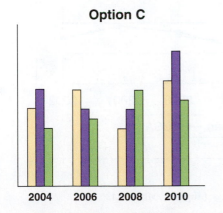

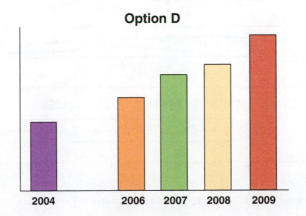

■ **Figure 3-9** ■ Bar chart variations

>> Bar Chart Guideline 4: Carefully Arrange the Order of Bars

The arrangement of bars is what reveals meaning to readers. Following are two common approaches:

- **Sequential:** Used when the progress of the bars shows a trend
- **Ascending or descending order:** Used when you want to make a point by the rising or falling of the bars

Line Charts

Line charts are a common graphic and work by using vertical and horizontal axes to reflect quantities of two different variables. The vertical (or *y*) axis usually plots the dependent variable; the horizontal (or *x*) axis usually plots the independent variable. (The dependent variable is affected by changes in the independent variable.) Lines then connect points that have been plotted on the chart.

>> Line Chart Guideline 1: Use Line Charts for Trends

Readers are affected by the direction and angle of the chart's line(s), so take advantage of this persuasive potential. In Figure 3-10, for example, the writer wants to show changes in gas prices over time. Including a line chart in the study gives immediate emphasis to the recent downward trend in book purchases.

Figure A9-3: Industrial Sector Gas Prices in the United States, OECD Europe, Japan, and Taiwan, 1994-2002, in 2003 Dollars

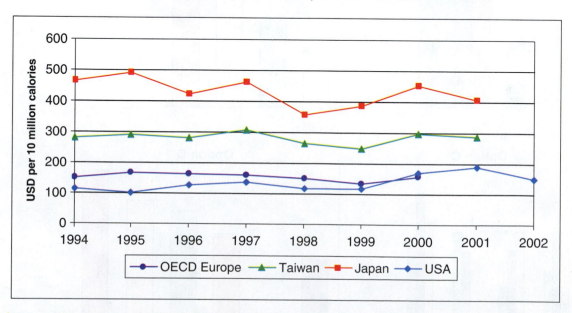

- **Figure 3-10** - Line chart using multiple lines

Source: The Economic Future of Nuclear Power: A Study Conducted at The University of Chicago (August 2004): A9–A13.

>> Line Chart Guideline 2: Locate Line Charts with Care

Given their strong impact, line charts can be especially useful as attention-grabbers. Consider placing them (1) on cover pages (to engage reader interest in the document), (2) at the beginning of sections that describe trends, and (3) in conclusions (to reinforce a major point of your document).

>> Line Chart Guideline 3: Strive for Accuracy and Clarity

Be sure that the line or lines on the graph truly reflect the data from which you have drawn. Also, select a scale that does not mislead readers with visual gimmicks. Following are some specific suggestions to keep line charts accurate and clear:

- Start all scales from zero to eliminate the possible confusion of breaks in amounts.
- Select a vertical-to-horizontal ratio for axis lengths that is pleasing to the eye (three vertical to four horizontal is common).
- Make chart lines as thick as, or thicker than, the axis lines.

>> Line Chart Guideline 4: Do Not Place Numbers on the Chart Itself

Line charts derive their main effect from the simplicity of lines that show trends. Avoid cluttering the chart with a lot of numbers that only detract from the visual impact.

>> Line Chart Guideline 5: Use Multiple Lines With Care

Like bar charts, line charts can show multiple trends. Simply add another line or two. To help readers quickly distinguish between lines, assign a differently shaped data point to each line. If you place too many lines on one chart, however, you run the risk of confusing the reader with too much data. Use no more than four or five lines on a single chart.

Flowcharts

Flowcharts tell a story about a process, usually by stringing together a series of boxes and other shapes that represent separate activities (Figure 3-11).

Some flowcharts use standardized symbols to represent steps in the decision-making process (Figure 3-12). Although these symbols were originally used for programming, they are now used to represent a wide range of processes. Because they have a reputation for being hard to read, you must take extra care in designing flowcharts.

mytechcommlab

See the flowchart in Model Brochure 1.

>> Flowchart Guideline 1: Present Only Overviews

Flowcharts should give only a capsule version of the process, not all details. Reserve your list of particulars for the text or the appendices, where readers expect it.

>> Flowchart Guideline 2: Limit the Number of Shapes

Flowcharts rely on rectangles and other shapes to tell a story. Different shapes represent different types of activities. Some flowcharts use icons and images to present information. This variety helps in describing a complex process, but it can also produce

■ **Figure 3-11** ■
Flowchart for basic
site survey project

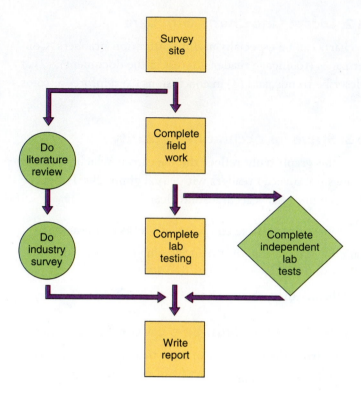

confusion. For the sake of clarity and simplicity, limit the number of different shapes in your flowcharts.

>> Flowchart Guideline 3: Provide a Legend When Necessary

Simple flowcharts often need no legend. The few shapes on the chart may already be labeled by their specific steps. When charts get more complex, however, include a legend that identifies the meaning of each shape used.

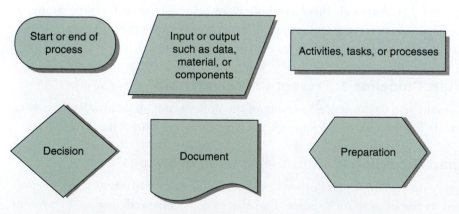

■ **Figure 3-12** ■ Selection of standard flowchart symbols

>> Flowchart Guideline 4: Run the Sequence from Top to Bottom or from Left to Right

Long flowcharts may cover the page with several columns or rows; however, they should always show some degree of uniformity by assuming either a basically vertical or horizontal direction.

>> Flowchart Guideline 5: Label All Shapes Clearly

Besides a legend that defines meanings of different shapes, the chart usually includes a label for each individual shape or step. Follow one of these approaches:

- Place the label inside the shape.
- Place the label immediately outside the shape.
- Put a number in each shape and place a legend for all numbers in another location (preferably on the same page).

Technical Drawings

Technical drawings are important tools that can accompany documents, such as instructions, reports, sales orders, proposals, brochures, and posters. They are preferred over photographs when specific views are more important than photographic detail.

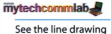

See the line drawing in Model Description 2.

>> Drawing Guideline 1: Choose the Right Amount of Detail

Keep drawings as simple as possible. Use only the level of detail that serves the purpose of your document and satisfies your readers' needs. For example, Figure 3-13 uses an exploded view to show a gear and its placement in a hybrid engine.

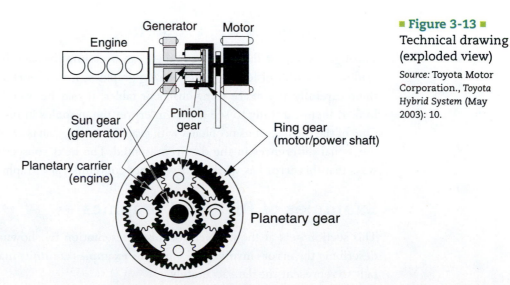

■ **Figure 3-13** ■
Technical drawing (exploded view)

Source: Toyota Motor Corporation., Toyota Hybrid System (May 2003): 10.

>> Drawing Guideline 2: Label Parts Well

Place labels on every part you want your reader to see. (Conversely, you can also choose not to label those parts that are irrelevant to your purpose.)

When you label parts, use a typeface large enough for easy reading. Also, arrange labels so that they are easy to locate and do not detract from the importance of the drawing itself.

>> Drawing Guideline 3: Choose the Most Appropriate View

Drawings offer you a number of options for perspective or view:

- **Exterior view** shows surface features with either a two- or three-dimensional appearance.
- **Cross-section view** shows a "slice" of the object so that interiors can be viewed.
- **Cutaway view** is similar to a cross section view, but only part of the exterior is removed to show the inner workings of the object.
- **Exploded view** shows relationship of parts to each other by "exploding" the mechanism.

See the cutaway drawing in Model Description 3: Flat Plate Solar Collector.

>> Drawing Guideline 4: Use Legends When There Are Many Parts

In complex drawings, avoid cluttering the illustration with many labels. Instead number parts and place labels with corresponding numbers in a list, or legend, at one side of the illustration.

>>> Misuse of Graphics

Technology has revolutionized the world of graphics by placing sophisticated tools in the hands of many writers. This largely positive event has a dark side in that many graphics distort data and misinform the reader. This final chapter section shows what can happen to graphics when sound design principles are not applied.

Description of the Problem

Through clutter or distortion, graphics can oversimplify data, be confusing, or be misleading. To avoid misleading or confusing graphics, it is important to proofread and edit them carefully. If you are using charts or tables, it may be useful to ask someone else to look at them carefully to see if they interpret the graphics in the way that you intended. One of the most serious problems with graphics is that charts are often disproportional to the actual differences in the data represented. The next subsection shows some specific ways that this error has worked its way into contemporary graphics.

Examples of Distorted Graphics

This section gets at the problem of misrepresentation by showing several examples and describing the errors involved. None of the examples commits major errors, yet each one fails to represent the data accurately.

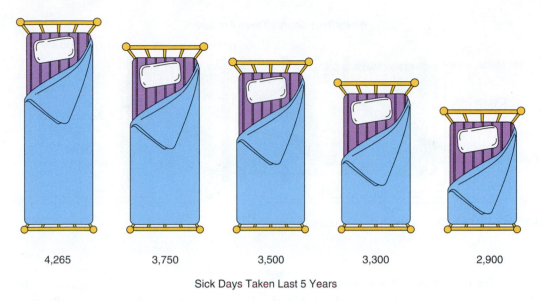

■ **Figure 3-14** ■
Faulty comparisons
on modified bar
chart

4,265   3,750   3,500   3,300   2,900

Sick Days Taken Last 5 Years

>> Example 1: Confusing Bar Charts

Figure 3-14 accompanied a report about sick days taken in an organization. The problem here is that the chart's decoration—the beds—inhibits rather than promotes clear communication. Although the writer intends to use bed symbolism in lieu of precise bars, the height of the beds does not correspond to the actual decrease in sick days. A revised graph appropriate for a report should include a traditional bar chart without the beds.

Another problem with bar charts is that they become busy or confusing, especially when the options for creating charts in spreadsheet programs are not used carefully. Figure 3-15 uses several stacked bars, which make it hard to compare the data clearly. It also uses too many colors, as well as colors that are similar, and that would not reproduce well on a black-and-white copier.

>> Example 2: Chartjunk That Confuses the Reader

Figure 3-16 concludes a report from a county government to its citizens. Whereas the dollar backdrop is meant to reinforce the topic—that is, the use to which tax funds are

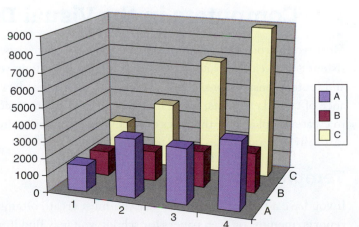

■ **Figure 3-15** ■
Confusing bar
chart

Where Your County Taxes Are Spent

Management 7%

Courts 24%

Police 18%

Bus service 13%

Community growth 4%

Parks and libraries 18%

Administrative services 11%

Other government 5%

■ **Figure 3-16** ■ *Chartjunk that confuses the reader*

put—in fact, it impedes communication. Readers cannot quickly see comparisons. Instead, they must read the entire list below the illustration, mentally rearranging the items into some order.

At the very least, the expenditures should have been placed in sequence, from least to greatest percentage or vice versa. Even with this order, however, one could argue that the dollar bill is a piece of what Edward Tufte, an expert in the visual display of data, calls *chartjunk*.

>> Example 3: Confusing Pie Charts

The pie chart in Figure 3-17 (1) omits percentages that should be attached to each of the budgetary expenditures; (2) fails to move in a largest-to-smallest clockwise sequence; (3) includes too many divisions, many of which are about the same size and thus difficult to distinguish; and (4) introduces a third dimension that adds no value to the graphic.

Pie charts should be perfect circles, should have percentages on the circle, and should move in large-to-small sequence from the 12:00 position.

>>> Computers in the Visual Design Process

Most word-processing programs include tools to make page design easier and more consistent and to easily create and insert graphics. They allow you to format running headers and footers, ensure consistency of elements such as headers and lists, change the appearance of tagged elements, insert images, and even create basic illustrations. They allow you to create templates for the types of documents that you write most often. They even automatically generate tables of contents and indexes.

Templates

If you have a type of document that you must create often, such as progress reports, lab reports, memos, or even papers for school, you may find it useful to create a template for

	FUND NAME	DESCRIPTION	FY '89 BUDGET
A	General Fund	Basic government activities	$112,895,822
B	Transit Fund	Implementation of bus system	10,812,522
C	Fire District Fund	Operation of Fire Department	21,253,523
D	Bond Funds	General obligation bond issue proceeds	11,073,371
E	Road Sales Tax Fund	1% special purpose sales tax for road improvements	116,869,904
F	Water & Pollution Control Fund	Daily water system operation	60,572,506
G	Debt Service Fund	Principal & interest payments for general obligation bonds	8,240,313
H	Water RE&I Fund	Maintenance of existing facilities	27,365,744
I	Solid Waste RE&I Fund	Maintenance of existing facilities	1,111,237
J	Solid Waste Disposal Facilities	Landfill operations	6,003,367
K	Water Construction Fund	Construction of new facilities	110,884,507
L	Other Uses*		18,384,364
		SUB-TOTAL	$505,467,180
		LESS INTERFUND ACTIVITY	– 23,393,042
		TOTAL EXPENDITURES	$482,074,138

*Other Uses includes: Community Service Block Grants, Law Library, Claims Fund, Capital Projects, Senior Services, Community Development Block Grant, Grant Fund

In addition to the General Fund, the county budgets a number of other specialized funds. These include the Fire District Fund, Transit Fund, Road Sales Tax Fund, and enterprise funds such as Water and Pollution Control, and Solid Waste Disposal Facilities.

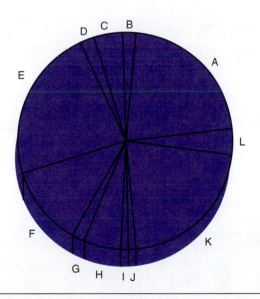

■ **Figure 3-17** ■ Confusing pie chart

Source: Adapted from Cobb County 1988–89 Annual Report (Cobb County, GA), 14. Used by permission.

that type of document. Your word processing program probably has several templates pre-loaded for memos, letters, and reports. Although these templates are handy, they may not exactly fit your needs. If you need to include a company logo on a letterhead or alter the headings in the report template, you can modify existing templates or you can create your own. Some software publishers also make a large number of templates available for downloading from their Web sites. Templates include passages of text or elements such as tables that are included in the same place in every document. They can also include a catalog of styles for elements, such as heading, lists, and even body text.

Style Sheets

When you are writing a long document with many headings or other typographical elements, it may be difficult to remember how you formatted each element. For example, if it has been several pages since you used a third-level heading, you may have to scroll back to see what type size you used and whether you bolded or italicized it. This problem can be solved by using the styles in your word processing software. A style sheet allows you to assign formatting to specific kinds of elements in your document, such as headings, body text, and lists. This formatting is done with *tags* or codes that your computer attaches to the elements. (If you are familiar with HTML coding, this tagging is similar.) You select the text, such as a first-level heading, select the appropriate style from a pull-down menu, and assign it to the selected text with a single mouse click. Figure 3-18 shows a style sheet, or *catalog,* that is part of a document template.

Running header with chapter title

Style catalog

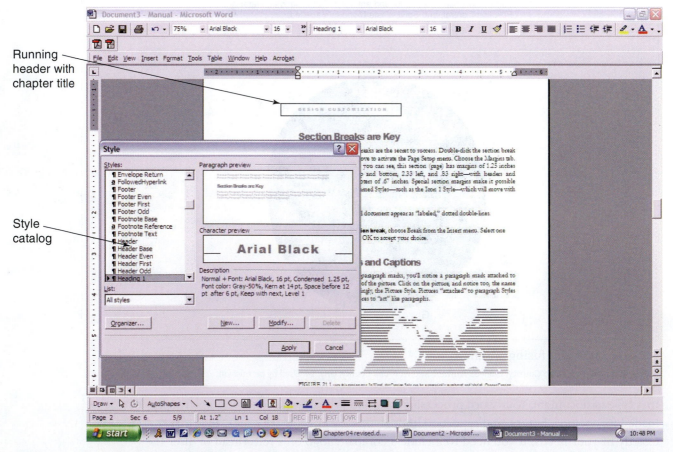

- **Figure 3-18** ■ Microsoft Word document template with style sheet

Source: "Microsoft product screen shot(s) reprinted with permission from Microsoft Corporation."

The heading tags that you created for your style sheet can also be used to automatically generate a Table of Contents. This process can be a bit complicated, so consult your program's Help file or an after-market manual for instructions about how to do this.

Learning to use the visual design tools in your word processing program can save you time and help you create consistent and professional-looking documents. However, these tools differ among the many word processing programs (and sometimes from one version of a word processing program to the next), so take the time to learn how to use the tools that are available in your word processing software.

>>> Chapter Summary

This chapter shows you how to apply principles of visual design to your assignments in your classes and your on-the-job writing. The term *visual design* refers to the array of design, layout, and graphic options you can use to improve the visual effect of your document.

Effective page design requires that you use specific elements, such as white space, headings, listings, and in-text emphasis. Another strategy for page design is to change the size and type of fonts in your documents. As with other strategies, this one must be used with care so that your document does not become too busy. Page design remains a technique for highlighting content, not a substitute for careful organization and editing.

More than ever before, readers of technical documents expect good graphics to accompany text, as well as special fonts and color. Graphics (also called *illustrations* or *visual aids*) can be in the form of (1) tables (rows and/or columns of data) or (2) figures (a catchall term for all non-table illustrations). Both tables and figures are used to simplify ideas, reinforce points made in the text, generate interest, and create a universal appeal.

The last several decades have seen an incredible change in the way documents are produced. Today, individual writers working at their personal computers write, edit, design, and print sophisticated documents. Writers who learn to use the visual design tools available in word processing programs find it easier to create consistent, professional-looking documents.

>>> Learning Portfolio

Collaboration at Work Critiquing an Annual Report

General Instructions

Each "Collaboration at Work" exercise applies strategies for working in teams to chapter topics. The exercise assumes you (1) have been divided into teams of about three to six students, (2) will use team time inside or outside of class to complete the case, and (3) will produce an oral or written response. For guidelines about writing in teams, refer to Chapter 2.

Background for Assignment

While planning and writing, you make two main decisions about the use of visual design elements—first, when they should be used; and second, what types to select. This chapter helps you make such decisions. Yet you already possess the quality that is most useful in your study of graphics: common sense. Whether consciously or subconsciously, most of us tend to seek answers to basic questions like the following when we read a document:

1. Are white space, headings, lists, font size and type, and color used to help readers find information?
2. Is there an appropriate mix of text and graphics?

3. Are the graphics really useful or are they just visual "fluff"?
4. Can information in the graphic be understood right away?
5. Was the correct type of graphic selected for the context?
6. Do any of the graphics include errors, such as in proportion?

Your answers to these questions often determine whether you continue reading a document—or at least whether you enjoy the experience.

Team Assignment

Locate a company's annual report in your library or choose another document that includes a variety of graphics—a newspaper, magazine, report, textbook, or catalog. Using the questions previously listed, work with your team to evaluate the use of graphics in all or part of the document. Whether you think a graphic is successful or not, give specific reasons to support your analysis.

Assignments

1. Document Navigation

Locate an example of technical writing that is at least ten pages long, such as a user's manual or instructions. Identify and analyze the navigation elements that the document uses to help users find information, including elements other than the three navigation tools discussed in this chapter. Your instructor will indicate whether your report should be oral or written.

2. Individual Practice in Page Design

As a manager at an engineering company, you have just finished a major report to a client. It gives recommendations for transporting a variety of hazardous materials by sea, land, and air. The body of your report contains a section that defines the term *stowage plan* and describes its use. Given your mixed technical and nontechnical audience, this basic information is much needed. What follows is the text of that

section. Revise the passage by applying any of this chapter's principles of page design that seem appropriate—such as adding headings, graphics, lists, and white space. If you wish, you may also make changes in organization and style. *Optional:* Share your version with another student to receive his or her response.

In the chemical shipping industry, a stowage plan is a kind of blueprint for a vessel. It lists all stowage tanks and provides information about tank volume, tank coating, stowed product, weight of product, loading port, and discharging port. A stowage plan is made out for each vessel on each voyage and records all chemicals loaded. The following information concerns cargo considerations (chemical properties and tank features) and some specific uses of the stowage plan in industry.

The three main cargo considerations in planning stowage are temperature, compatibility, and safety. Chemicals

In the chemical shipping industry, a stowage plan is a kind of blueprint for a vessel. It lists all stowage tanks and provides information about tank volume, tank coating, stowed product, weight of product, loading port, and discharging port. A stowage plan is made out for each vessel on each voyage and records all chemicals loaded. The following information concerns cargo considerations (chemical properties and tank features) and some specific uses of the stowage plan in industry.

The three main cargo considerations in planning stowage are temperature, compatibility, and safety. Chemicals have physical properties that distinguish them from one another. To maintain the natural state of chemicals and to prevent alteration of their physical properties, a controlled environment becomes necessary. Some chemicals, for example, require firm temperature controls to maintain their physical characteristics and degree of viscosity (thickness) and to prevent contamination of the chemicals by any moisture in the tanks. In addition, some chemicals, like acids, react violently with each other and should not be stowed in adjoining, or even neighboring, tanks. In shipping, this relationship is known as *chemical compatibility*.

The controlled environment and compatibility of chemicals have resulted in safety regulations for the handling and transporting of these chemicals. These regulations originate with the federal government, which bases them on research done by the private manufacturers. Location and size of tanks also determine the placement of cargo. A ship's tanks are arranged with all smaller tanks around the periphery of the tank grouping and all larger tanks in the center. These tanks, made of heavy steel and coated with zinc or epoxy, are highly resistant to most chemicals, thereby reducing the chance of cargo contamination. Each tank has a maximum cargo capacity, and the amounts of each chemical are matched with the tanks. Often chemicals to be discharged at the same port are staggered in the stowage plan layout so that after they are discharged the ship maintains its equilibrium.

The stowage plan is finalized after considering the cargo and tank characteristics. In its final form, the plan is used as a reference document with all information relevant to the loading/discharging voyage recorded. If an accident occurs involving a ship, or when questions arise involving discharging operations, this document serves as a visual reference and brings about quick decisions.

have physical properties that distinguish them from one another. To maintain the natural state of chemicals and to prevent alteration of their physical properties, a controlled environment becomes necessary. Some chemicals, for example, require firm temperature controls to maintain their physical characteristics and degree of viscosity (thickness) and to prevent contamination of the chemicals by any moisture in the tanks. In addition, some chemicals, like acids, react violently with each other and should not be stowed in adjoining, or even neighboring, tanks. In shipping, this relationship is known as *chemical compatibility*.

The controlled environment and compatibility of chemicals have resulted in safety regulations for the handling and transporting of these chemicals. These regulations originate with the federal government, which bases them on research done by the private manufacturers. Location and size of tanks also determine the placement of cargo. A ship's tanks are arranged with all smaller tanks around the periphery of the tank grouping and all larger tanks in the center. These tanks, made of heavy steel and coated with zinc or epoxy, are highly resistant to most chemicals, thereby reducing the chance of cargo contamination. Each tank has a maximum cargo capacity, and the amounts of each chemical are matched with the tanks.

Often chemicals to be discharged at the same port are staggered in the stowage plan layout so that after they are discharged the ship maintains its equilibrium.

The stowage plan is finalized after considering the cargo and tank characteristics. In its final form, the plan is used as a reference document with all information relevant to the loading/discharging voyage recorded. If an accident occurs involving a ship, or when questions arise involving discharging operations, this document serves as a visual reference and brings about quick decisions.

3. Team Practice in Page Design: Using Computer Communication

This assignment is feasible only if you and your classmates have access to software that allows you to post messages to team members, edit on screen, and send edited copy back and forth. Your task is to add appropriate page design features to either (a) the *stowage plan* excerpt in Assignment 2 or (b) any other piece of unformatted text permitted for use by your instructor. Choose a team leader who will collect and collate the individual edits. Choose another team member to type or scan the excerpt into the computer and then e-mail the passage to other team members. Then each person should add the features desired and e-mail the edited document to the team leader, who will collate the revisions and e-mail the new version to team members for a final edit. Throughout this process, participants may conduct e-mail conversations about the draft and resolve differences, if possible, before sending drafts to the leader. The team may need one or two short meetings in person, but most business should be conducted via the computer. The goal is to arrive at one final version for your team.

4. Pie, Bar, and Line Charts

Figure 3-19 shows employment by industry from 2000 through 2006, while also breaking down the 2006 data into four categories by race. Use those data to complete the following charts:

- A pie chart that shows the groupings of race in 2006.
- A bar chart that shows the trend in total employment during 2000, 2004, 2005, and 2006. Indicate the gap in data.

- A segmented bar chart that compares employment in the production of durable goods to employment in the production of nondurable goods within the manufacturing sector for 2004, 2005, and 2006.
- A single-line chart showing employment in agriculture and related industries for 2000 through 2006.
- A multiple-line chart that contrasts employment in retail trade, professional and business services, and leisure and hospitality for 2004 through 2006.

5. Flowcharts

Identify the main activities involved in enrolling in classes on your campus. Then draw two flowcharts that outline the main activities involved in this process. In the first chart, use the standard flowchart symbols shown in Figure 3-12 on page 62. In the second flowchart, use images and symbols creatively to explain the process.

Table 602. Employment by Industry: 2000 to 2006

[**In thousands (136,891 represents 136,891,000), except percent.** See Table 584 regarding coverage and headnote Table 587 regarding industries]

Industry	2000	2004 [1]	2005 [1]	2006 [1]	2006, percent [1]			
					Female	Black [2]	Asian [2]	Hispanic [3]
Total employed	136,891	139,252	141,730	144,427	46.3	10.9	4.5	13.6
Agriculture and related industries	2,464	2,232	2,197	2,206	24.6	2.7	1.2	19.4
Mining	475	539	624	687	13.0	4.9	0.7	13.6
Construction	9,931	10,768	11,197	11,749	9.6	5.5	1.4	25.1
Manufacturing	19,644	16,484	16,253	16,377	29.5	9.5	5.2	14.7
Durable goods	12,519	10,329	10,333	10,499	25.8	8.5	5.8	12.4
Nondurable goods	7,125	6,155	5,919	5,877	36.1	11.4	4.2	18.7
Wholesale trade	4,216	4,600	4,579	4,561	29.0	6.5	4.1	13.5
Retail trade	15,763	16,269	16,825	16,767	48.9	10.1	4.2	12.7
Transportation and utilities	7,380	7,013	7,360	7,455	24.2	16.5	3.6	12.7
Transportation and warehousing.	6,096	5,844	6,184	6,269	24.7	17.6	3.8	13.5
Utilities	1,284	1,168	1,176	1,186	21.9	10.9	2.5	8.2
Information	4,059	3,463	3,402	3,573	44.4	11.7	5.2	9.4
Financial activities.	9,374	9,969	10,203	10,490	55.5	10.2	5.1	10.0
Finance and insurance	6,641	6,940	7,035	7,254	58.2	10.5	5.6	8.5
Real estate and rental and leasing . . .	2,734	3,029	3,168	3,237	49.4	9.5	4.1	13.4
Professional and business services	13,649	14,108	14,294	14,868	42.5	9.8	5.7	13.0
Professional and technical services. . .	8,266	8,386	8,584	8,776	44.4	6.4	7.6	6.2
Management, administrative, and waste services	5,383	5,722	5,709	6,092	39.8	14.8	3.0	22.9
Education and health services.	26,188	28,719	29,174	29,938	74.9	14.2	4.7	9.1
Educational services	11,255	12,058	12,264	12,522	68.9	10.8	3.6	8.5
Health care and social assistance. . . .	14,933	16,661	16,910	17,416	79.1	16.7	5.4	9.5
Hospitals	5,202	5,700	5,719	5,712	76.6	16.4	7.0	7.6
Health services, except hospitals . .	7,009	8,118	8,332	8,639	78.6	15.3	5.3	9.5
Social assistance	2,722	2,844	2,860	3,065	85.4	21.2	2.9	12.9
Leisure and hospitality	11,186	11,820	12,071	12,145	51.3	10.5	5.9	19.4
Arts, entertainment, and recreation . . .	2,539	2,690	2,765	2,671	45.2	8.3	3.6	11.9
Accommodation and food services . . .	8,647	9,131	9,306	9,474	53.0	11.2	6.5	21.6
Other services	6,450	6,903	7,020	7,088	51.7	9.8	5.8	15.5
Other services, except private households.	5,731	6,124	6,208	6,285	46.5	9.6	6.2	13.3
Private households	718	779	812	803	92.5	11.1	2.5	32.8
Government workers	6,113	6,365	6,530	6,524	45.4	16.2	3.5	8.6

[1] See footnote 2, Table 569.   [2] Persons in this race group only. See footnote 3, Table 570.   [3] Persons of Hispanic or Latino origin may be of any race.

Source: U.S. Bureau of Labor Statistics, *Employment and Earnings,* monthly, January 2007 issue. See Internet site <http://www.bls.gov/cps/home.htm>.

■ **Figure 3-19** ■ Reference for Assignment 4

Source: U.S. Census Bureau. (2008). *The 2008 Statistical Abstract.* www.census.gov/compendia/statab/2008/tables/08s0602.xls.

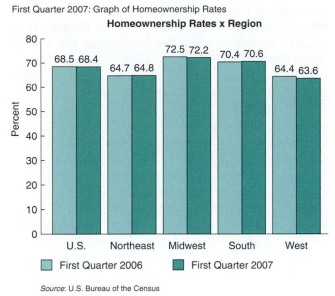

First Quarter 2007: Graph of Homeownership Rates

Homeownership Rates x Region

Source: U.S. Bureau of the Census

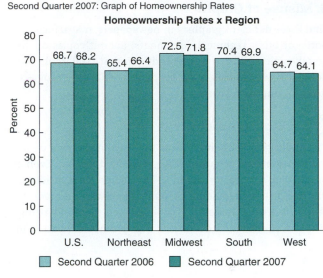

Second Quarter 2007: Graph of Homeownership Rates

Homeownership Rates x Region

Source: U.S. Bureau of the Census

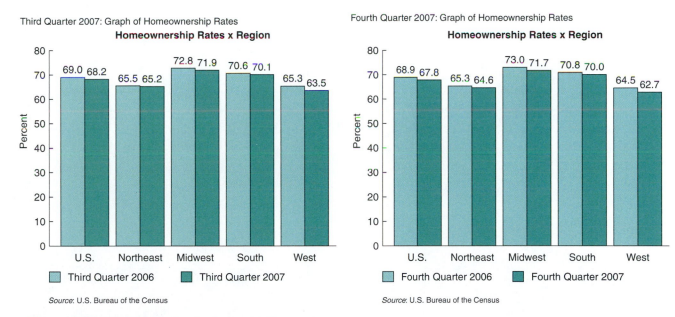

Third Quarter 2007: Graph of Homeownership Rates

Homeownership Rates x Region

Source: U.S. Bureau of the Census

Fourth Quarter 2007: Graph of Homeownership Rates

Homeownership Rates x Region

Source: U.S. Bureau of the Census

■ **Figure 3-20** ■ Reference for Assignment 7

Source: U.S. Census Bureau. (2008). *Housing Vacancies and Homeownership.* http://www.census.gov/hhes/www/housing/hvs/hvs.html

6. Technical Drawing

Drawing freehand, using the draw function in your word processor, or using a computer illustration or design program available to you, produce a simple technical drawing of an object with which you are familiar through work, school, or home use.

7. Table

Using the bar charts in Figure 3-20, create a table that shows home ownership by region and total home ownership for the United States for each quarter of 2007.

8. Misuse of Graphics

Find three deficient graphics in newspapers, magazines, reports, or other technical documents. Submit copies of the graphics along with a written critique that (1) describes in detail the deficiencies of the graphics and (2) offers suggestions for improving them.

? 9. Ethics Assignment

Visual design greatly influences the way people read documents no matter what message is being delivered. Even a product, a service, or an idea that could be considered harmful to the individual or public good can be made to seem more acceptable by a well-designed piece of writing. Find a well-designed document that promotes what is, in your opinion, a harmful product, service, or idea. Explain why you think the writers made the design decisions they did. Your example should include page design and graphic design elements.

10. International Communication Assignment

Collect one or more samples of business or technical writing that originate in—or are designed for—cultures outside the United States. (Use either print examples or examples found on the Internet.) Comment on features of visual design in the samples. If applicable, indicate how such features differ from those evident in business and technical writing designed for an audience within the United States.

Chapter | **4** | **Letters, Memos, and Electronic Communication**

>>>Chapter Outline

You may write more letters, memos, and e-mails in your career than any other type of document. Generally termed "correspondence," letters, memos, and e-mails are short documents written to accomplish a limited purpose. Letters are directed outside your organization, memos are directed within your organization, and e-mail can be directed to either an external or an internal audience.

Your ability to write good memos, letters, and e-mails depends on a clear sense of purpose, understanding of reader needs, and close attention to formats. This chapter prepares you for this challenge by presenting sections that cover general rules that apply to all workplace correspondence and specific formats for positive correspondence, negative correspondence, and neutral correspondence. Also included is advice specific to letter, memoranda, and e-mail formats. Job letters and resumes are discussed in a separate chapter on the job search (Chapter 10).

>>> General Guidelines for Correspondence

Letters convey your message to readers outside your organization, just as memos are an effective way to get things done within your own organization, and e-mail is a way to communicate quickly with readers inside and outside of your organization. By applying the guidelines in this chapter, you can master the craft of writing effective correspondence. Refer to the Figure 4-2 on page 83, Figure 4-3 on page 84, and Figure 4-4 on page 87 in this chapter for examples that demonstrate the guidelines that follow.

>> Correspondence Guideline 1: Know Your Purpose

Before beginning your draft, write down your purpose in one clear sentence. This *purpose sentence* often becomes one of the first sentences in the document. Following are some samples:

■ **Letter purpose sentence:** "As you requested yesterday, I'm sending samples of the new candy brands you are considering placing your office vending machines."

■ **Memo purpose sentence:** "This memo explains the organization's new policy for selecting rental cars on business trips."

■ **E-mail purpose sentence:** "I have attached the most recent draft of the proposal for the PI Corp. pipeline project."

Some purpose statements are implied; others are stated. An implied purpose statement occurs in the first paragraph of Figure 4-2. That paragraph indicates that the letter is a response to an earlier letter that requested a change in the construction schedule. Figure 4-3 shows a more obvious purpose statement in the third sentence. The third sentence in Figure 4-4 tells the reader exactly what will follow.

>> Correspondence Guideline 2: Know Your Readers

Who are you trying to inform or influence? Pay particular attention when correspondence will be read by more than one person. If these readers are from different technical levels or different administrative levels within an organization, the challenge increases. A complex audience compels you to either (1) reduce the level of technicality to that which can be understood by all readers or (2) write different parts of the document for different readers.

>> Correspondence Guideline 3: Follow Correct Format

Most organizations adopt letter and memo formats that must be used uniformly by all employees. Later in this chapter, we offer specific guidelines for formatting letters, memos, and e-mail, but the following are some general formatting guidelines:

- **Multiple-page headings:** Each page after the first page of a letter or memo often has a heading that includes the name of the person or company receiving the letter or memo, the date, and the page number. Some organizations may prefer an abbreviated form such as "Jones to Bingham, 2," without the date.

- **Subject line:** Memos and e-mails always include subject lines, but they are sometimes used in letters as well. Give the subject line special attention, because it telegraphs meaning to the audience immediately. In fact, readers use it to decide when, or if, they will read the complete correspondence. Be brief, but also engage interest. For example, the subject line of the Figure 4-2 memo could have been "Changes." Yet, that brevity would have sacrificed reader interest. The actual subject line, "Copy Center Changes," conveys more information and shows readers that the contents of the memo will make their lives easier.

- **Enclosure notation:** If attachments or enclosures accompany a letter or memo, type the singular or plural form of "Enclosure" or "Attachment" one or two lines beneath the reference initials. Some writers also list the item itself (e.g., Enclosure: Code of Ethics). If an e-mail includes an attachment, the filename and file type (e.g. PDF) should be included in the first paragraph.

- **Copy notation:** If the correspondence has been sent to anyone other than the recipient, type "Copy" or "Copies" one or two lines beneath the enclosure notation, followed by the name(s) of the person or persons receiving copies (e.g., Copy: Preston Hinkley). E-mail inserts this information automatically. If you are sending a copy but do not want the original letter or memo to include a reference to that copy, write "bc" (for *blind copy*) and the person's name only on the copy—not on the original (e.g., bc: Mark Garibaldi). (Note: Send blind copies only when you are certain it is appropriate and ethical to do so.)

>> Correspondence Guideline 4: Follow the ABC Format for All Correspondence

Correspondence subscribes to the same three-part ABC (Abstract/Body/Conclusion) format used throughout this book. According to the ABC format, your correspondence is composed of these three main sections:

- **Abstract:** The abstract introduces the purpose and usually gives a summary of main points to follow. It includes one or two short paragraphs.

mytechcommlab

Internal Information was written for a specific audience.

■ **Body:** The body contains supporting details and thus makes up the largest part of a letter or memo. You can help your readers by using such techniques as:

Deductive patterns for paragraphs: In this general-to-specific plan, your first sentence should state the point that helps the reader understand the rest of the paragraph. This pattern avoids burying important points in the middle or end of the paragraph, where they might be missed. Fast readers tend to focus on paragraph beginnings and expect to find crucial information there. Note how most paragraphs in Figure 4-3 follow this format.

Personal names: If they know you, readers like to see their names in the body of the letter or memo, or in the salutation of an e-mail. Your effort here shows concern for the reader's perspective, gives the correspondence a personal touch, and helps strengthen your personal relationship with the reader. (See the last paragraph in Figure 4-2.) Of course, the same technique can sometimes backfire, because it is an obvious ploy to create an artificially personal relationship.

Lists that break up the text: Listed points are a good strategy for highlighting details. Readers are especially attracted to groupings of three items, which create a certain rhythm, attract attention, and encourage recall. Use bullets, numbers, dashes, or other typographic techniques to signal the listed items. For example, the bulleted list in Figure 4-3 draws attention to three important points about that the writer wants to emphasize. Because some e-mail systems cannot read special characters like bullets, use asterisks or dashes for lists in e-mail.

Strongest points first or last: If your correspondence presents support or makes an argument, include the most important points at the beginning or at the end—not in the middle. For example, the first paragraph in Figure 4-3 announces the changes in the Copy Center, and the last paragraph tells readers when the changes will take place.

Headings to divide information: One-page letters and memos, and even e-mail, sometimes benefit from the emphasis achieved by headings. The three headings in Figure 4-3 quickly steer the reader to main parts of the document.

■ **Conclusion:** Readers remember first what they read last. The final paragraph of your correspondence should leave the reader with an important piece of information— for example, (1) a summary of the main idea or (2) a clear statement of what will happen next. The Figure 4-1 letter promises a successful outcome, whereas the Figure 4-4 e-mail ends with a reminder that the sender has submitted a proposal to the recipient.

>> Correspondence Guideline 5: Use the 3Cs Strategy for Persuasive Messages

The ABC format provides a way to organize all letters and memos. Another pattern of organization for you to use is the *3Cs strategy*—especially when your correspondence has a persuasive objective. This strategy has three main goals:

■ **Capture** the reader's interest with a good opener, which tells the reader what the letter, memo, or e-mail can do for him or her.

- **Convince** the reader with supporting points, all of which confirm the opening point that this document will make life easier.
- **Contact** solidifies your relationship with the reader with an offer to follow up on the correspondence.

Note, how each of the samples in this chapter uses the 3Cs strategy.

>> Correspondence Guideline 6: Stress the "You" Attitude

As noted earlier, using the reader's name in the body helps convey interest. However, your efforts to see things from the reader's perspective must go deeper than a name reference. For example, you should perform the following tasks:

- Anticipate questions your reader might raise and then answer these questions. You can even follow an actual question ("And how will our new testing lab help your firm?") with an answer ("Now TestCorp's labs can process samples in 24 hours.").
- Replace the pronouns *I, me,* and *we* with *you* and *your*. Of course, you must use first-person pronouns at certain points in a letter, but many pronouns should be second person. The technique is quite simple. You can change almost any sentence from writer-focused prose ("We feel that this new service will . . . ") to reader-focused prose ("You'll find that this new service will . . . ").

Figure 4-2 shows this *you* attitude by emphasizing how the sender's organization can help the recipient with the scheduling change. Figure 4-4 shows it by responding directly to the sender's concerns.

>> Correspondence Guideline 7: Use Attachments for Details

Keep text brief by placing details in attachments, which readers can examine later, rather than bogging down the middle of the letter or memo.

>> Correspondence Guideline 8: Be Diplomatic

Without a tactful tone, all your planning and drafting will be wasted. Choose words that persuade and cajole, not demand. Be especially careful of memos written to subordinates. If you sound too authoritarian, your message may be ignored—even if it is clear that what you are suggesting will help the readers. Generally speaking, negative (or "bad news") letters often use the passive voice, whereas positive (or "good news") letters often use the active voice.

For example, the letter in Figure 4-2 would fail in its purpose if it flatly refused to make changes and offered no alternatives.

>> Correspondence Guideline 9: Respond Quickly

A letter, memo, or e-mail that comes too late fails in its purpose, no matter how well written. Mail letters within 48 hours of your contact with, or request from, the reader. Send memos in plenty of time for your reader to make the appropriate adjustments in schedule, behavior, and so forth. Usually respond to e-mails the same day you receive them.

mytechcommlab

Correspondence can be broadly categorized as positive, negative, and neutral. For examples of correspondence that meets specific purposes, see the Model letters, memos, and e-mail.

Claim Response is an
effective example of
a positive letter.

Positive Correspondence

Everyone likes to give good news; fortunately, you will often be in the position of providing it when you write. Following are some sample situations:

- Replying to a question about products or services
- Recommending a colleague for a promotion or job
- Responding favorably to a complaint or an adjustment

The trick is to recognize the good-news potential of many situations. This section gives you an all-purpose format for positive correspondence.

ABC Format: Positive Correspondence

- **ABSTRACT:** Bridge between this correspondence and last communication with person
  - Clear statement of good news you have to report
- **BODY:** Supporting data for main point mentioned in abstract
  - Clarification of any questions reader may have
  - Qualification, if any, of the good news
- **CONCLUSION:** Statement of eagerness to continue relationship, complete project, etc.
  - Clear statement, if appropriate, of what step should come next

ABC Format for Positive Correspondence

All positive correspondence follows one overriding rule. You must always:

State good news immediately!

Any delay gives readers the chance to wonder whether the news will be good or bad, thus causing momentary confusion. On the left is a complete outline for positive letters that corresponds to the ABC format.

Negative Correspondence

Complaint is a
carefully written
negative letter.

It would be nice if all your correspondence could be as positive as that just described. Unfortunately, the real world does not work like that. You will have many opportunities to display both tact and clarity in relating negative information. Following are a few cases:

- Explaining delays in projects or delivery of services
- Registering complaints about products or services
- Giving bad news about employment or performance

This section gives you a format to follow in writing sensitive correspondence with negative information.

ABC Format: Negative Correspondence

- **ABSTRACT:** Bridge between your correspondence and previous communication
  - General statement of purpose or appreciation—in an effort to find common bond or area of agreement
- **BODY:** Strong emphasis on what can be done, when possible
  - Buffered yet clear statement of what cannot be done, with clear statement of reasons for negative news
  - Facts that support your views
- **CONCLUSION:** Closing remarks that express interest in continued association
  - Statement, if appropriate, of what will happen next

ABC Format for Negative Correspondence

One main rule applies to all negative correspondence:

Buffer the bad news, but still be clear.

Despite the bad news, you want to keep the reader's goodwill. Spend time at the beginning building your relationship with the reader by introducing less controversial information—before you zero in on the main message. On the right is an overall pattern to apply in all negative correspondence.

Neutral Correspondence

Some correspondence expresses neither positive nor negative news. It is simply the routine correspondence written every day to keep businesses and other organizations operating. Some situations follow:

- Requesting information about a product or service
- Inviting the reader to an event
- Sending solicited or unsolicited items through the mail

Use the following outline in writing your neutral letters.

mytechcommlab
Memo 10: Web site Request is a good example of a neutral memo.

ABC Format for Neutral Correspondence

Because the reader usually has no personal stake in the news, neutral correspondence requires less emphasis on tone and tact than other types, yet it still requires careful planning. In particular, always abide by this main rule:

Be absolutely clear about your inquiry or response.

Neutral correspondence operates a bit like positive correspondence. You must make your point early, without giving the reader time to wonder about your message. Neutral correspondence varies greatly in specific organization patterns. The *umbrella plan* suggested here emphasizes the main criterion of clarity.

ABC Format: Neutral Correspondence

- **ABSTRACT:** Bridge or transition between correspondence and previous communication, if any
  - Precise purpose of correspondence (e.g., request, invitation, response to invitation)
- **BODY:** Details that support the purpose statement (e.g., a description of item(s) requested, the requirements related to the invitation, a description of item(s) being sent)
- **CONCLUSION:** Statement of appreciation
  - Description of actions that should occur next

>>> Letters

Letters are to your clients and vendors what memos are to your colleagues. They relay information quickly and keep business flowing. Here is a working definition:

Letter: A document that conveys information to a member of one organization from someone outside that same organization. Letters usually cover one major point and fit on one page. This chapter classifies letters into these three groups, according to type of message: (1) positive, (2) negative, and (3) neutral.

■ **Figure 4-1** ■ Block
and modified block
style for letters

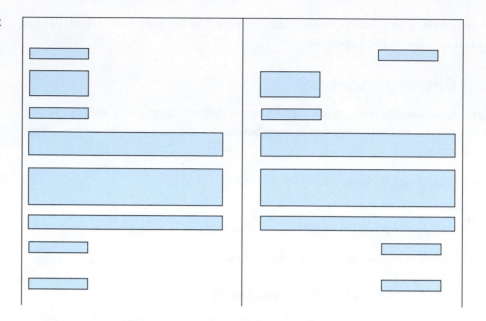

There are two main letter formats—block and modified block. Figure 4-1 shows the basic page design of each; Figure 4-2 is in block style. As noted, you usually follow the preferred format of your own organization. Whatever format you choose, most or all business letters include the following sections: address of the sender (usually in letterhead), date, address of the receiver, greeting, body, signature, and reference to any attachments and/or those individuals being copied.

Addresses on envelopes and in letters should use the format recommended by the United States Postal Service. Addresses should include no more than four lines, and should not include punctuation such as commas or periods.

>>> Memoranda

Even though e-mail has become common in the workplace, *memoranda* (the plural of *memorandum,* also called *memos*) are still important. Even if you work in an organization that uses e-mail extensively, you will still compose print messages that convey your point with brevity, clarity, and tact. Here is a working definition:

Memorandum: A document written from a member of an organization to one or more members of the same organization. Abbreviated *memo*, it usually covers just one main point and no more than a few. Readers prefer one-page memos.

With minor variations, all memos look much the same. The obligatory "Date/To/From/Subject" information hangs at the top left margin, in whatever order your organization requires. Figure 4-3 shows one basic format. These four lines allow you to dispense with lengthy introductory passages seen in more formal documents. Note that the sender signs his or her initials after or above the typed name in the "From" line.

12 Post Street

Houston Texas 77000

713.555.1381

July 23, 2011

The Reverend Mr John C Davidson

Maxwell Street Church

Canyon Valley

Texas 79195

Dear Reverend Davidson:

Thanks for your letter asking to reschedule the foundation project at your church from mid-August to late August, because of the regional conference. I am sure you are proud that Maxwell was chosen as the conference site.

One reason for our original schedule, as you may recall, was to save the travel costs for a project crew going back and forth between Houston and Canyon Valley. Because **Delvientos Construction** has several other jobs in the area, we had planned not to charge you for travel.

We can reschedule the project, as you request, to a more convenient date in late August, but the change will increase project costs from $1,500 to $1,800 to cover travel. At this point, we just do not have any other projects scheduled in your area in late August that would help defray the additional expenses. Given our low profit margin on such jobs, that additional $300 would make the difference between our firm making or losing money on the foundation repair at your church.

I'll call you next week, Reverend Davidson, to select a new date that would be most suitable. **Delvientos Construction** welcomes its association with the Maxwell Street Church and looks forward to a successful project in late August.

Sincerely,

Nancy Slade

Project Manager

NS/mh

File #34678

Implied purpose response to previous letter.

Opens on positive note.

Reminds him about original agreement—in tactful manner

Phrases negative message as positively as possible, giving rationale for necessary change.

Closes with promise of future contact.

Makes it clear what will happen next. Ends on positive note.

Writer's initials/typists initials.

■ **Figure 4-2** ■ Negative letter in block style

Memorandum

Date: August 1, 2011

To: All Employees

From: Gini Preston
          Chair
          Copy Services Committee

Subject: Copy Center Changes

Gives brief purpose statement and overview of contents. →

With the purchase of two new copiers and a mechanical folder, the Copy Center is able to expand its services. At the same time, we have had to reduce the paper stock that we keep on hand because of space limitations. This memo highlights the services and products now available at the Copy Center.

Color copies

With our new equipment, color copies do not require additional time to process. However, because color copies are expensive, please limit your use of them. If you have a document that includes both color and black-and-white pages, submit them as separate jobs so that the color copier is used only for color copies.

Special stock

The Copy Center now stocks only two colors of paper in addition to white paper: blue and goldenrod. Cover stock is available only in white and blue. We continue to stock transparencies. Although we are no longer stocking other kinds of paper, we are still able to meet requests for most special stock:

Emphasizes need for special handling of requests for special paper. →

- **Stocks available with 24-hour notice:** We can purchase 11 x 17 inch paper, cover stock and regular stock in a variety of colors, and specialized paper such as certificates and NCR (carbonless copy) paper. Departments will be charged for all special stock.
- **Coated stock:** Our copiers do not produce quality copies on coated stock (paper or cover stock with a slick coating, like magazine paper). We will continue to outsource jobs that use coated stock to KDH Printing. Please allow at least one week for jobs that use coated stock.

Bindery services

With our new equipment, collating and stapling of large jobs no longer require additional time. The following bindery services are also available in house, but may require additional time:

- Perfect and spiral binding

- Folding

Makes it clear when changes will take place. →

- Cutting and hole punching. (The paper cutter and paper drill can be used on up to 500 sheets at a time.)

The new equipment will be available August 15. Your efforts to make the most efficient use of Copy Center resources help improve the quality of your documents and the productivity of the company.

Invites contact. →

Feel free to call me at ext. 567 if you have any questions.

■ **Figure 4-3** ■ Memorandum: changes in services

Abide by this one main rule in every memo-writing situation:

Be clear, brief, and tactful.

Because many activities are competing for their time, readers expect information to be related as quickly and clearly as possible; however, you must be sure not to sacrifice tact and sensitivity as you strive to achieve conciseness.

>>> E-Mail

Electronic communication (*e-mail*) has become the preferred means of communication for many people in their professional lives.

E-mail is an appropriate reflection of the speed at which we conduct business today. Following are some of the obvious advantages that using e-mail provides:

- It gets to the intended receiver quickly.
- Its arrival can be confirmed easily.
- Your reader can reply to your message quickly.
- It's cheap to use—once you have invested in the hardware and software.
- It permits cheap transmission of multiple copies and attachments. Any medium so widely used deserves special attention in a chapter on correspondence. Here is a working definition:

> **E-mail: A document written often in an informal style either to members of one's own organization or to an external audience. E-mail messages often cover one main point. Characterized by the speed with which it is written and delivered, an e-mail can include more formal attachments to be read and possibly printed by the audience.**

mytechcommlab

Workplace E-mail is an effective, neutral e-mail.

Adding to the ease of transmission is the fact that e-mail allows you to create mailing lists. One address label can be an umbrella for multiple recipients, saving you much time.

Of course, the flip side of this ease of use: E-mail is *not* private. Every time you send an e-mail, remember that it may be archived or forwarded, and may end up being read by "the world." Either by mistake or design, many supposedly private e-mails often are received by unintended readers.

Appropriate Use and Style for E-mail

Simply understanding that e-mail *should* have a format puts you ahead of many writers, who consider e-mail a license to ramble on without structure. Yes, e-mail is casual and quick, but that does not make it formless. Remember, the configurations of some computers make reading a screen harder on the eyes than reading print memos.

So give each e-mail a structure that makes it simple for your reader to find important information. Computers and e-mail systems handle formatting of texts and special characters differently, so you should format your e-mail so that it can be read on any computer. Use your system's default font, and avoid highlighting, color, bold, italics, and underlining. E-mails are generally short, no longer than what can be seen on a computer screen all at once, and paragraphs should be short. Some e-mail systems cannot translate tabs, so use lines of white space between paragraphs. Figure 4-4 shows how one writer clearly separates topics within an e-mail.

Guidelines for E-mail

When writing e-mail, you should try to strike a balance between speed of delivery on the one hand and quality of the communication to your reader on the other hand. In fact, the overriding rule for e-mail is as follows:

Do not send it too quickly!

By taking an extra minute to check the style and tone of your message, you have the best chance of sending an e-mail that will be well received.

>> E-mail Guideline 1: Use Style Appropriate to the Reader and Subject

E-mail sent early in a relationship with a client or other professional contact should be somewhat formal. It should be written more like a letter, with a salutation, closing, and complete sentences. E-mail written once a professional relationship has been established can use a more casual style. It can resemble conversation with the recipient on the phone. Sentence fragments and slang are acceptable, as long as they contribute to your objectives and are in good taste. Most important, avoid displaying a negative or angry tone. Do not push the *Send* button unless an e-mail produces a constructive exchange.

>> E-mail Guideline 2: Be Sure Your Message Indicates the Context to Which It Applies

Tell your readers what the subject is and what prompts you to write your message. If you are replying to a message, be sure to include the previous message or summarize the message to which you are replying. Most e-mail software packages include a copy of the message to which you are replying. However, you should make sure that you include only the messages that provide the context for your reader. Long strings of forwarded e-mail make it difficult to find the necessary information.

```
**************************************************
```

X-Sender: mckinley@mail2.AlvaCon.com

Date: Tue, Nov 11, 2008 09:25:30 -0800

To: pcarmich@advantage.com

From: Mike McKinley <mckinley@mail2.AlvaCon.com>

Subject: Our Recent Visit

Mime-Version: 1.0

Dear Paul,

I enjoyed meeting you also and visiting with your staff. I particularly enjoyed meeting Harold Black, for he will be very valuable in developing the plans for the possible water purification plant.

My responses to your concerns are below.

YOU WROTE:

>If Advantage, Inc., does decide to build the water purification plant, we would be very interested in having Delvientos Construction's Mary Stevens as the project manager.

REPLY:

That certainly will be a possibility; Mary is one of our best managers.

YOU WROTE:

>After you left, I called the city administration here in Murrayville. Delvientos Construction does not need a business license for your work here, but, of course, you will need the necessary construction permits.

REPLY:

Thanks for taking care of this matter—I had not thought of that. We will supply the details to you for applying for the construction permits if you accept our proposal.

```
**************************************************
```

Positive opening, with reference to specific events.

Clearly indicates purpose: to respond to an earlier e-mail.

Clear indication of original e-mail text and of reply.

Indicates next action.

Closes with references to pending proposal.

■ **Figure 4-4** ■ An e-mail message that separates different topics for reply

>> **E-mail Guideline 3:** Choose the Most Appropriate Method for Replying to a Message

Short e-mail messages may only require that you include a brief response at the beginning or end of the e-mail to which you are responding. For complex, multi-topic messages, however, you may wish to split your reply by commenting on each point individually (Figure 4-4).

>> **E-mail Guideline 4:** Format Your Message Carefully

Because e-mail messages frequently replace more formal print-based documents, they should be organized and formatted so that the readers can locate the information you want to communicate easily.

- Use headings to identify important chunks of information.
- Use lists to display a series of information.
- Use sufficient white space to separate important chunks of information.
- Use separators to divide one piece of information from another.

>> **E-mail Guideline 5:** Chunk Information for Easy Scanning

Break the information into coherent chunks dealing with one specific topic, including all the details that a reader needs to get all of the essential information. Depending on the nature of the information, include specific topic, time, date, location, and necessary prerequisites and details.

>> **E-mail Guideline 6:** When Writing to Groups, Give Readers a Method to Abstain from Receiving Future Notices

E-mail can easily become invasive and troublesome for recipients. You will gain favor—or at least not lose favor—if you are considerate and allow recipients to decide what e-mail they wish to receive.

>> **E-mail Guideline 7:** When Writing to Groups, Suppress the E-mail Addresses of Recipients—Unless the Group Has Agreed to Let Addresses Be Known

It is inappropriate to reveal the e-mail addresses of group members to other group members. Use the "BC" line to suppress group members' addresses.

>> **E-mail Guideline 8:** When Composing an Important Message, Consider Composing It in Your Word Processor

Important e-mail messages should be not only clear in format, but also correct in mechanics. Because e-mail software may not have a spelling checker, compose important messages in your word processor and use your spelling checker to check accuracy. Then either cut and paste it into an e-mail message or attach it as a file.

E-mail communication is often considered less formal and, therefore, less demanding in its format and structure than print-based messages, such as memoranda and letters. However, because e-mail messages have become so pervasive a means of communication, you should consider constructing them as carefully as you would a memorandum or a letter.

Another reason to exercise great care is that e-mail, like conventional documents, can be used in legal proceedings and other formal contexts.

>>> Memoranda Versus E-mail

Although e-mail has become the most common form of internal correspondence in the workplace, there are times when a memo is a better option. Send a memo instead of an e-mail in the following situations:

- The document is longer than can be viewed easily on a computer screen.
- The document must include symbols, special characters, or other formatting that may not be available through all computer systems.
- The document includes graphics.
- The document must be posted in print form.
- The document contains sensitive information, including information about clients, projects, or personnel.

>>> Chapter Summary

Correspondence keeps the machinery of business, industry, and government moving. Letters usually are sent to readers outside your organization, whereas memos are sent to readers inside. E-mail can be sent to internal or external audiences. In all types of correspondence, abide by these rules:

1. Know your purpose.
2. Know your readers.
3. Follow correct format.
4. Follow the ABC format for all letters and memos.
5. Use the 3Cs strategy for persuasive messages.
6. Stress the *you* attitude.
7. Use attachments for details.
8. Be diplomatic.
9. Edit carefully.

Besides following these basics, you must follow specific strategies for positive, negative, and neutral correspondence. In correspondence with a positive message, the good news always goes first. In correspondence with a negative message, work on maintaining goodwill by placing a buffer before the bad news. Neutral correspondence, such as requests for information, should be absolutely clear in its message. Letters develop and maintain professional relationships with those outside of your organization. Memos should strive for brevity, clarity, and tact. Your relationship with both superiors and subordinates can depend in part on how well you write memoranda. E-mail messages give you the additional flexibility of adopting an informal and more conversational style. Use e-mail when speed and informality are desired.

>>> Learning Portfolio

Collaboration at Work Choosing the Right Mode

General Instructions

Each Collaboration at Work exercise applies strategies for working in teams to chapter topics. The exercise assumes you (1) have been divided into teams of about three to six students, (2) use team time inside or outside of class to complete the case, and (3) produce an oral or written response. For guidelines about writing in teams, refer to Chapter 2.

Background for Assignment

A century ago, business professionals had few opportunities for communication beyond the formal letter or meeting; today, the range of options is incredibly broad. On the one hand, we marvel at the choices for getting our message heard or read; on the other hand, the many ways to communicate present an embarrassment of riches that can be confusing.

In other words, when you have multiple communication options, you are challenged to match the right method with the right context—*right* in terms of what the reader wants and *right* in terms of the level of effort you should

exert to suit the purpose. You may think this challenge applies only to your working life. However, it also can influence your life in college, as this exercise shows.

Team Assignment

Brainstorm with your team to list every means you have used to communicate with your college and university, from the time you applied to the present. Then for each communication option that follows, provide two or three situations for which the option is the appropriate choice:

1. Letter that includes praise
2. Letter that describes a complaint
3. Letter that provides information
4. Letter that attempts to persuade
5. Telephone call
6. Fax transmission
7. E-mail
8. Memorandum
9. Personal meeting

Assignments

Follow these general guidelines for these assignments:

- Print or design a letterhead when necessary.
- Use whatever letter format your instructor requires.
- Invent addresses when necessary.
- Invent any extra information you may need for the correspondence, but do not change the information presented here.

1. Positive Letter—Favorable Response to Complaint

The following letter was written in response to a complaint from an office manager. She wrote to the manufacturer that the lunchroom microwave broke down just three days after the warranty expired. Although she did not ask for a specific monetary adjustment, she did make clear her extreme dissatisfaction with the product. The manufacturer responded with the following letter. Be prepared to discuss what is right and what is wrong with the letter. Also, rewrite it using this chapter's guidelines.

This letter is in response to your August 3 complaint about the Justrite microwave oven you purchased about six months ago for your lunchroom at Zocalo Realty, Inc. We understand that the turntable in the microwave broke shortly after the warranty expired.

Did you know that last year our microwave oven was rated "best in its class" and "most reliable" by *Consumers Count* magazine? Indeed, we have received so few complaints about the product that a recent survey of selected purchasers revealed that 98.5% of first-time purchasers of our microwave ovens are pleased that they chose our products and would buy another.

Please double-check your microwave to make sure that the turntable is broken—it may just be temporarily stuck. We rarely have had customers make this specific complaint about our product. However, if the turntable is in need of repair, return the entire appliance to us, and we will have it repaired free of charge or have a new replacement sent to you. We stand behind our product, because the warranty period only recently expired.

It is our sincere hope that you continue to be a satisfied customer of Justrite appliances.

2. Negative Letter—Explanation of Project Delay

You work for Delvientos Construction. As project manager for the construction of a small strip shopping center, you have had delays about halfway through the project because of bad weather. Even worse, the forecast is for another week of heavy rain. Yesterday, just when you thought nothing else could go wrong, you discovered that your concrete supplier, Atlas Concrete, has a truck drivers' strike in progress. Because you still need half the concrete for the project, you have started searching for another supplier.

mytechcommlab
For practice in organization, do the Regency Real Estate Memo Activity.

Your client, an investor/developer named Tanya Lee located in a city about 200 miles away, probably will be upset by any delays in construction, whether or not they are within your control. Write her a letter in which you explain weather and concrete problems. Try to ease her concern, especially because you want additional jobs from her in the future.

3. Negative Letter—Change in Project Scope and Schedule

As a marketing executive at an engineering firm, you oversee many of the large accounts held by the office. One important account is a company that owns and operates a dozen radio and television stations throughout the Midwest. On one recent project, your organization's engineers and technicians did the foundation investigation for, and supervised construction of, a new transmitting tower for a television station in Toledo. First, your staff members completed a foundation investigation, at which time they examined the soils and rock below grade at the site. On the basis of what they learned, you ordered the tower and the guy wires that connect it to the ground. Once the construction crew actually began excavating for the foundation, however, they found mud that could not support the foundation for the tower. Although unfortunate, it sometimes happens that actual soil conditions cannot be predicted by the preliminary study. Because of this discovery of mud, the tower must be shifted to another location on the site. As a result, the precut guy wires are the wrong length for the new site, requiring you to order wire extenders. The extenders will arrive in two weeks, delaying placement of the tower by that much time. All other parts of the project are on schedule, so far.

Your client, Ms. Sharon West of Midwest Media Systems in Cleveland, does not understand much about soils and foundation work, but she does understand what construction delays mean to the profit margin of her firm's new tele-

vision station as it attempts to compete with larger stations in Toledo. You must console this important client while informing her of this recent finding.

4. Neutral Letter—Response to Request for Information

As reservations clerk for the Best Central Inn in St. Louis, you just received a letter from Jerald Pelletier, an administrative secretary with MovieStream, making arrangements for a meeting of shipping center managers from around the country. The group is considering holding its quarterly meeting in St. Louis in six months. Pelletier has asked you to send some brief information on hotel rates, conference facilities (meeting rooms), and availability. Send him room rates for double and single rooms, and let him know that you have four conference rooms to rent out at $150 each per day. Also, tell him that at this time, the hotel rooms and conference rooms are available for the three days he mentioned.

5. Memo—Negative News

You are regional manager of Allen Security, a national security firm. The current policy in your organization states that employees must pass a pre-employment drug screening before being hired. After that, there are no tests unless you or one of your job supervisors has reason to suspect that an employee is under the influence of drugs on the job.

mytechcommlab
For practice in appropriate style, do the Revising an Adjustment Letter Activity.

Lately, a number of clients have strongly suggested that you should have a random drug-screening policy for all employees in the on-site personnel group. They argue that the on-the-job risk to life and property is great enough to justify this periodic testing. You have consulted your branch managers, who like the idea. You have also talked with the company's attorney, who assures you that such random testing should be legal, given the character of the group's work. After considerable thought, you decide to implement the policy in three weeks. Write a memo to all employees of your group and relate this news.

6. Memo—Persuasive Message

For this assignment, choose either (1) a good reference book or textbook in your field of study or (2) an excellent periodical in your field. The book or periodical should be one that could be useful to someone working in a profession, preferably one that you may want to enter.

Now assume that you are an employee of an organization that would benefit by having this book or periodical in its staff library, customer waiting room, or perhaps as a reference book purchased for employees in your group. Write a one-page memo to your supervisor recommending the purchase. You might want to consider criteria, such as:

- Relevance of information in the source to the job
- Level of material with respect to potential readers
- Cost of book or periodical as compared with its value
- Amount of probable use
- Important features of the book or periodical (such as bibliographies or special sections)

7. E-mail—Positive News

As Human Resources Director of a large theme park developer, you just learned from your accounting firm that last year's profits were even higher than previously expected. Apparently, proceeds from your company's newest water park had not been counted in the first reporting of profits. You and your managers had already announced individual raises before you learned this good news. Now you want to write an e-mail that states that every employee will receive a $500 across-the-board bonus, in addition to whatever individual raises have been announced for next year. Include the subject line for the e-mail.

8. E-mail—Neutral Message

As a mailroom supervisor, you have a number of changes to announce to employees of the corporate office. Write an e-mail, including the subject line that clearly relates the following information: Deliveries and pickups of mail, which currently are at 8:30 A.M. and 3:00 P.M., will change to 9:00 A.M. and 3:30 P.M., starting in two weeks. Also, there will be an additional pickup at noon on Monday, Wednesday, and Friday. The mailroom will start picking up mail to go out by Federal Express or any other one-day carrier, rather than the sender having to wait for the carrier's representative to come to the sender's office. The sender must call the mailroom to request the pickup; and the carrier must be told by the sender to go to the mailroom to pick up the package. The memo should also remind employees that the mail does not go out on federal holidays, even though the mailroom continues to pick up mail from the offices on those days.

9. Ethics Assignment

Pooling the experience that members of your team have had with e-mail, focus specifically on inappropriate or unethical behavior. Possible topics include the content of messages, tone of language, and the use of distribution lists. Now draft a simple e-mail Code of Ethics that could be distributed to members of any organization whose members use e-mail on a daily basis. To find examples of actual codes of ethics, use "code of ethics" in an Internet search.

10. International Communication Assignment

E-mail messages can be sent around the world as easily as they can be sent to the next office. If you end up working for a company with international offices or clients, you probably will use e-mail to conduct business.

Investigate the e-mail conventions of one or more countries outside your own. Search for any ways that the format, content, or style of international e-mail may differ from e-mail in your country. Gather information by collecting hard copy of e-mail messages sent from other countries, interviewing people who use international e-mail, or consulting the library for information on international business communication. Write a memo to your instructor in which you (1) note differences you found and (2) explain why these differences exist. If possible, focus on any differences in culture that may affect e-mail transactions.

Chapter 5 | Definitions and Descriptions

efinitions, descriptions, process explanations, and instructions are the types of writing that people often think of when they think of technical communication. This chapter and Chapter 6 cover these four elements of technical communication.

Definitions and descriptions are closely related; in fact, descriptions often begin with a definition. Process explanations and instructions are also closely related, with the difference being how the reader will use the documents.

mytechcommlab

Description 5 opens with a formal definition sentence.

>>> Definitions Versus Descriptions

During your career, you will use technical terms known only to those in your profession. As a civil engineer, for example, you would know that a *triaxial compression test* helps determine the strength of soil samples. As a documentation specialist, you would know that *single-sourcing* allows the creation of multiple documents from the same original text. When writing to readers unfamiliar with these fields, however, you must define technical terms. You may also have to describe technical objects, and the distinction between *definition* and *description* can sometimes be a bit confusing.

Definitions and descriptions can appear in any part of a document, from the introduction to the appendix, or they may be created as stand-alone documents. Good definitions can support findings, conclusions, and recommendations throughout your document. They also keep readers interested. Conversely, the most organized and well-written report falls on deaf ears if it includes terms that readers do not grasp. For your reader's sake, then, you must be asking questions like these about definitions:

- How often do you use them?
- Where should they be placed?
- What format should they take?
- How much information is enough, and how much is too much?

To answer these questions, the following sections give guidelines for definitions and supply annotated examples.

Descriptions are similar to definitions. In fact, they often open with a short definition, but they also emphasize the physical details of the object being described. Like definitions, descriptions often appear as supporting information in the document body or in appendices.

>>> Guidelines for Writing Definitions

Once you know definitions are needed, you must decide on their format and location. Again, consider your readers. How much information do they need? Where is this information best placed within the document? To answer these and other questions, we offer five working guidelines for writing good definitions.

>> Definition Guideline 1: Keep It Simple

Although the sole purpose of a report occasionally is to define a term; most often a definition is used to clarify a term in a document with a larger purpose. Definitions should be as simple and unobtrusive as possible, with only that level of detail needed by the reader.

Choose from the following three main formats (listed from least to most complex) in deciding the form and length of definitions:

- **Informal definition:** A word or brief phrase, often in parentheses, that gives only a synonym or other minimal information about the term.

- **Formal definition:** A full sentence that distinguishes the term from other similar terms and includes these three parts: the term itself, a class to which the term belongs, and distinguishing features of the term.

- **Expanded definition:** A lengthy explanation that begins with a formal definition and is developed into several paragraphs or more.

Guidelines 2–5 show you when to use these three options and where to put them in your document.

>> Definition Guideline 2: Use Informal Definitions for Simple Terms Most Readers Understand

Informal definitions appear right after the terms being defined, often as one-word synonyms in parentheses. They give just enough information to keep the reader moving quickly and are best used with simple terms that can be defined without much detail.

>> Definition Guideline 3: Use Formal Definitions for More Complex Terms

A formal definition, like the one in Figure 5–1, appears in the form of a sentence that lists (1) the *term* to be defined, (2) the *class* to which it belongs, and (3) the *features* that distinguish the term from others in the same class. Use it when your reader needs more background than an informal definition provides. Formal definitions define in two stages:

- First, they place the term into a *class* (group) of similar items.

- Second, they list *features* (characteristics) of the term that separate it from all others in that same class.

In the list of sample definitions that follows, note that some terms are intangible (like *arrest*) and others are tangible (like *pumper*). Yet, all can be defined by first choosing a class of objects or concepts and then selecting features that distinguish the term from others in the same class.

Term	Class	Features
An arrest is	restraint of persons	that deprives them of freedom of movement and binds them to the will and control of the arresting officer.
A *financial statement* is	a historical report about a business	prepared by an accountant to provide information useful in making economic decisions, particularly for owners and creditors.
A *triaxial compression test* is	a soils lab test	that determines the amount of force needed to cause a shear failure in a soil sample.
A *pumper*	is a fire-fighting apparatus	used to provide adequate pressure to propel streams of water toward a fire.

This list demonstrates three important points about formal definitions. First, the definition itself must not contain terms that are confusing to your readers. The definition of *triaxial compression test,* for example, assumes readers understand the term *shear failure* that is used to describe features. If this assumption is incorrect, then the term must be defined. Second, formal definitions may be so long that they create a major distraction in the text. (See Guideline 5 for alternative locations.) Third, the class must be narrow enough so that you do not have to list too many distinguishing features.

Formal sentence definition with term class, and features.

BMI [Body Mass Index] is a practical measure that requires only two things: accurate measures of an individual's weight and height (Figure 1). BMI is a measure of weight in relation to height. BMI is calculated as weight in pounds divided by the square of the height in inches, multiplied by 703. Alternatively, BMI can be calculated as weight in kilograms divided by the square of the height in meters.

Excerpted from The Surgeon General's call to action to prevent and decrease overweight and obesity. (2001). Office of Disease Prevention and Health Promotion. Centers for Disease Control and Preventions, National Institutes of Health. Rockville, MD: U.S. Department of Health and Human Services, Public Health Service, Office of the Surgeon General. Washington.

■ **Figure 5–1** ■ Example definition with formal sentence definition

>> Definition Guideline 4: Use the ABC Format for Expanded Definitions

Sometimes a parenthetical phrase or formal sentence definition is not enough. If readers need more information, a definition can be expanded to a paragraph, a page, or even multiple pages in length. Expanded definitions like the one in Figure 5–2 use this three-part structure:

- The **Abstract** component provides an overview at the beginning, including a formal sentence definition and a description of the ways you will expand the definition.

- The **Body** component provides supporting information using headings and lists as helpful format devices for the reader

- The **Conclusion** component should be brief, reminding the reader of the definition's relevance to the whole document.

The NIST Definition of Cloud Computing
Authors: Peter Mell and Tim Grance
Version 15, 10-7-09
National Institute of Standards and Technology, Information Technology Laboratory

Note 1: Cloud computing is still an evolving paradigm. Its definitions, use cases, underlying technologies, issues, risks, and benefits will be refined in a spirited debate by the public and private sectors. These definitions, attributes, and characteristics will evolve and change over time.

Note 2: The cloud computing industry represents a large ecosystem of many models, vendors, and market niches. This definition attempts to encompass all of the various cloud approaches.

Definition of Cloud Computing:
Cloud computing is a model for enabling convenient, on-demand network access to a shared pool of configurable computing resources (e.g., networks, servers, storage, applications, and services) that can be rapidly provisioned and released with minimal management effort or service provider interaction. This cloud model promotes availability and is composed of five essential **characteristics,** three **service models**, and four **deployment models**. ← *Overview, including formal sentence definition of term*

Essential Characteristics:
On-demand self-service. A consumer can unilaterally provision computing capabilities, such as server time and network storage, as needed automatically without requiring human interaction with each service's provider.
Broad network access. Capabilities are available over the network and accessed through standard mechanisms that promote use by heterogeneous thin or thick client platforms (e.g., mobile phones, laptops, and PDAs).

■ **Figure 5–2** ■ Expanded definition

Source: National Institute of Standards and Technology Computer Security Division, Computer Security Resource Center. http://csrc.nist.gov/groups/SNS/cloud-computing/cloud-def-v15.doc.

List of
components

Resource pooling. The provider's computing resources are pooled to serve multiple consumers using a multi-tenant model, with different physical and virtual resources dynamically assigned and reassigned according to consumer demand. There is a sense of location independence in that the customer generally has no control or knowledge over the exact location of the provided resources but may be able to specify location at a higher level of abstraction (e.g., country, state, or datacenter). Examples of resources include storage, processing, memory, network bandwidth, and virtual machines.

Rapid elasticity. Capabilities can be rapidly and elastically provisioned, in some cases automatically, to quickly scale out and rapidly released to quickly scale in. To the consumer, the capabilities available for provisioning often appear to be unlimited and can be purchased in any quantity at any time.

Measured Service. Cloud systems automatically control and optimize resource use by leveraging a metering capability at some level of abstraction appropriate to the type of service (e.g., storage, processing, bandwidth, and active user accounts). Resource usage can be monitored, controlled, and reported providing transparency for both the provider and consumer of the utilized service.

Service Models:

Cloud Software as a Service (SaaS). The capability provided to the consumer is to use the provider's applications running on a cloud infrastructure. The applications are accessible from various client devices through a thin client interface such as a web browser (e.g., web-based email). The consumer does not manage or control the underlying cloud infrastructure including network, servers, operating systems, storage, or even individual application capabilities, with the possible exception of limited user-specific application configuration settings.

Cloud Platform as a Service (PaaS). The capability provided to the consumer is to deploy onto the cloud infrastructure consumer-created or acquired applications created using programming languages and tools supported by the provider. The consumer does not manage or control the underlying cloud infrastructure including network, servers, operating systems, or storage, but has control over the deployed applications and possibly application hosting environment configurations.

Information
about context

Cloud Infrastructure as a Service (IaaS). The capability provided to the consumer is to provision processing, storage, networks, and other fundamental computing resources where the consumer is able to deploy and run arbitrary software, which can include operating systems and applications. The consumer does not manage or control

■ **Figure 5–2** ■ continued

the underlying cloud infrastructure but has control over operating systems, storage, deployed applications, and possibly limited control of select networking components (e.g., host firewalls).

Deployment Models:

Private cloud. The cloud infrastructure is operated solely for an organization. It may be managed by the organization or a third party and may exist on premise or off premise.

Community cloud. The cloud infrastructure is shared by several organizations and supports a specific community that has shared concerns (e.g., mission, security requirements, policy, and compliance considerations). It may be managed by the organizations or a third party and may exist on premise or off premise.

Public cloud. The cloud infrastructure is made available to the general public or a large industry group and is owned by an organization selling cloud services.

Hybrid cloud. The cloud infrastructure is a composition of two or more clouds (private, community, or public) that remain unique entities but are bound together by standardized or proprietary technology that enables data and application portability (e.g., cloud bursting for load-balancing between clouds).

Note: Cloud software takes full advantage of the cloud paradigm by being service oriented with a focus on statelessness, low coupling, modularity, and semantic interoperability.

Information about context

Summary of value of model

■ **Figure 5–2** ■ continued

Following are seven ways to expand a definition:

1. Background or history of term
2. Explanation of how the term is applied in context
3. List of parts
4. Graphics
5. Comparison/contrast of familiar or related terms
6. Explanation of underlying basic principles
7. Example

mytechcommlab

For another example of an expanded definition, see the Model Extended Definition: Diabetes Circular.

>> Definition Guideline 5: Choose the Right Location for Your Definition

Short definitions are likely to be in the main text; long ones are often relegated to footnotes or appendices. However, length is not the main consideration. Think first about the *importance* of the definition to your reader. If you know that decision makers reading your report need the definition, then place it in the text—even if it is fairly lengthy. If the definition provides only supplementary information, then it can go elsewhere. You have these five choices for locating a definition:

1. **In the same sentence as the term,** as with an informal, parenthetical definition.

2. **In a separate sentence,** as with a formal sentence definition occurring right after a term is mentioned.

3. **In a footnote,** as with a formal listed at the bottom of the page on which the term is first mentioned.

4. **In a glossary at the beginning or end of the document,** along with all other terms needing definition in that document.

5. **In an appendix at the end of the document,** as with an expanded definition that would otherwise clutter the text of the document.

>>> Guidelines for Writing Descriptions

When readers will benefit from detailed information about parts, functions, or other elements, you should write a description (Figure 5–3). These five guidelines help you write accurate, detailed descriptions. Follow them carefully as you prepare assignments in this class and on the job.

>> Description Guideline 1: Remember Your Readers' Needs

The level of detail in a technical description depends on the purpose a description serves. Give readers precisely what they need—no more. Always know just how much detail will get the job done.

>> Description Guideline 2: Be Accurate and Objective

More than anything else, readers expect accuracy in descriptions. Pay close attention to details. Along with accuracy should come *objectivity*. This term is more difficult to pin down, however. Some writers assume that an objective description leaves out all opinion. This is not the case. Instead, an objective description may very well include opinions that have these features:

- They are based on your professional background.
- They can be justified by the time you have had to complete the description.
- They can be supported by details from the site or object being described.

>> Description Guideline 3: Choose an Overall Organization Plan

Technical descriptions usually make up only parts of documents. Nevertheless, they must have an organization plan that permits them to be read as self-contained, stand-alone sections. Indeed, a description may be excerpted later for separate use.

The front panel

Your printer's front panel is located on the front of the printer, on the right hand side. Use it for the following functions:

- Use it to perform certain operations, such as loading and unloading paper.
- View up-to-date information about the status of the printer, the ink cartridges, the printheads, the maintenance cartridge, the paper, the print jobs, and other parts and processes.
- Get guidance in using the printer.
- See warning and error messages, when appropriate.
- Use it to change the values of printer settings and the operation of the printer. However, settings in the Embedded Web Server or in the driver override changes made on the front panel.

Starts with overview of important functions.

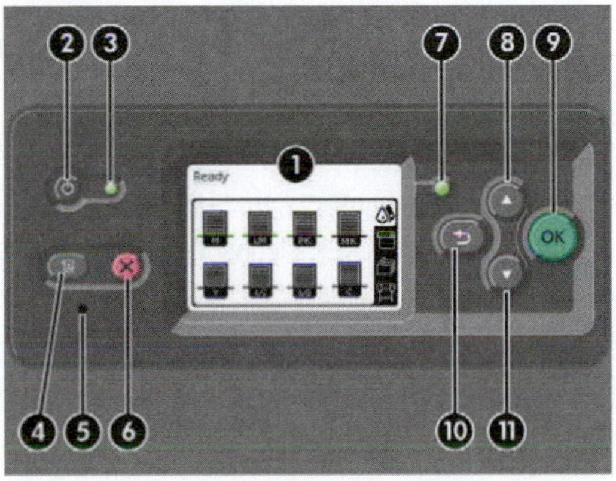

Illustration focuses on parts being described

The front panel has the following components:

1. The display area shows information, icons, and menus.
2. The Power button turns the printer on and off. If the printer is in sleep mode, this button will wake it up. (This is different from the hard power switch on the back of the printer. See Turn the printer on and off on page 21.)
3. The Power light is off when the printer is off. This light is amber when the printer is in sleep mode, green when the printer is on, green and flashing when the printer is in transition between off and on.
4. The Form Feed and Cut button normally advances and cuts the roll. Here is a list of its other functions:
   - If the printer is waiting for more pages to be nested, this button cancels the waiting time and prints the available pages immediately.
   - If the printer is drying the ink after printing, this button cancels the waiting time and releases the page immediately.
   - If the take-up reel is enabled, this button advances the paper 10 cm (3.9 inches), but does not cut the paper.

Numbers correspond to parts in illustration.

Integrates description of parts with operating instructions.

■ **Figure 5–3** ■ Description from a user's manual

(Content courtesy of Hewlett-Packard Company)

5. The Reset button restarts the printer (as if it were switched off and switched on again). You will need a nonconductive implement with a narrow tip to operate the Reset button.
6. The Cancel button cancels the current operation. It is often used to stop the current print job.
7. The Status light is off when the printer is not ready to print: the printer is either off, or in sleep mode. The Status light is green when the printer is ready and idle, green and flashing when the printer is busy, amber when a serious internal error has occurred, and amber and flashing when the printer is awaiting human attention.
8. The UP button moves to the previous item in a list, or increases a numerical value.
9. The OK button is used to select the item that is currently highlighted.
10. The Back button is used to return to the previous menu. If you press it repeatedly, or hold it down, you return to the main menu.
11. The Down button moves to the next item in a list, or decreases a numerical value.

To *highlight* an item on the front panel, press the Up or Down button until the item is highlighted.

To *select* an item on the front panel, first highlight it and then press the OK button.

The four front-panel icons are all found on the main menu. If you need to select or highlight an icon, and you do not see the icons in the front panel, press the Back button until you can see them.

Sometimes this guide shows a series of front panel items like this: **Item1 > Item2 > Item3**. A construction like this indicates that you should select **Item1,** select **Item2,** and then select **Item3.**

Refers user to more detailed information. → You will find information about specific uses of the front panel throughout this guide.

▪ Figure 5–3 ▪ continued

Following are three common ways to describe physical objects and events. In all three cases, a description should move from general to specific—that is, you begin with a view of the entire object or event, and in the rest of the description, you focus on specifics. Headings may be used, depending on the format of the larger document.

1. **Description of the parts:** For many physical objects you simply organize the description by moving spatially from part to part.
2. **Description of the functions:** Often the most appropriate overall plan relies on how things work, not on how they look.
3. **Description of the sequence:** If your description involves events, you can organize ideas around the major actions that occurred, in their correct sequence. As with any list, it is best to place a series of many activities into just a few groups. It is much easier for readers to comprehend four groups of five events each than a single list of 20 events.

mytechcommlab

Description 4 is well organized, although, as noted, some of its formatting could be improved.

>> Description Guideline 4: Use "Helpers" Like Graphics and Analogies

The words of a technical description must come alive. Because your readers may be unfamiliar with the item, you must search for ways to connect with their experience and with their senses. Two effective tools are graphics and analogies.

Graphics respond to the desire of most readers to see pictures along with words. As readers move through your part-by-part or functional breakdown of a mechanism, they can refer to your graphic aid for assistance. The illustration helps you too, of course, in that you need not be as detailed in describing locations and dimensions of parts when you know the reader has easy access to a visual.

Analogies, like illustrations, give readers a convenient handle for understanding your description. Put simply, an analogy allows you to describe something unknown or uncommon in terms of something that is known or more common. A brief analogy can sometimes save you hundreds of words of technical description, especially in a description of a concept.

>> Description Guideline 5: Give Your Description the "Visualizing Test"

After completing a description, test its effectiveness by reading it to someone unfamiliar with the material—someone with about the same level of knowledge as your intended reader. If this person can draw a rough sketch of the object or events while listening to your description, then you have done a good job. If not, ask your listener for suggestions to improve the description.

>>> Chapter Summary

Definitions and descriptions help readers who are unfamiliar with technical terms understand your documents and use the documents you write to make informed decisions.

Definitions occur in technical communication in one of three forms: *informal* (in parentheses), *formal* (in sentence form with term, class, and features), and *expanded* (in a paragraph or more). The following main guidelines apply:

1. Keep it simple.
2. Use informal definitions for simple terms most readers understand.
3. Use formal definitions for more complex terms.
4. Use the ABC Format for expanded definitions.
5. Choose the right location for your definition.

Descriptions, like definitions, depend on detail and accuracy for their effect. Careful descriptions usually include a lengthy itemizing of the parts of a mechanism or the functions of a term. Follow these basic guidelines for producing effective descriptions:

1. Remember your readers' needs.
2. Be accurate and objective.
3. Choose an overall organization plan.
4. Use "helpers" like graphics and analogies.
5. Give your description the "visualizing test."

>>> Learning Portfolio

Collaboration at Work Analyzing the Core

General Instructions

Each Collaboration at Work exercise applies strategies for working in teams to chapter topics. The exercise assumes you (1) have been divided into teams of about three to six students, (2) will use team time inside or outside of class to complete the case, and (3) will produce an oral or written response. For guidelines about writing in teams, refer to Chapter 2.

Background for Assignment

Whereas some terms are easily defined, others, including abstract concepts, can be quite challenging. This also means that it is essential that abstract terms be clearly defined, because not all readers will understand the term the way that you do. This assignment asks your team to define the abstract concept of a college or university education.

Like other colleges and universities, your institution may require students to complete a core curriculum of required subjects. Some cores are virtually identical for students in all majors; others vary by major. Following is one example of a core curriculum (required for institutions in the University System of Georgia):

1. **Essential Skills (9 semester hr):** Includes two freshman composition courses and college algebra.
2. **Institutional Options (4–5 semester hr):** Includes courses of the institution's choosing, such as public speaking and interdisciplinary classes.
3. **Humanities/Fine Arts (10–11 semester hr):** Includes courses, such as literature surveys, art appreciation, music appreciation, and foreign language.
4. **Science, Mathematics, and Technology (10–11 semester hr):** Includes laboratory and non-laboratory classes in fields, such as physics, chemistry, biology, and calculus.
5. **Social Sciences (12 semester hr):** Includes courses, such as American history, world history, political science, and religion.
6. **Courses Related to Student's Program of Study (18 semester hr):** Includes lower-level classes related to the student's specific major. For example, an engineering major may be required to take extra math, whereas a technical communication major may be required to take introductory technical communication.

Core curricula or general studies requirements like these suggest a definition of a college or university education. In this example, the required courses suggest that the university values developing intellectual curiosity and an understanding of communication, critical thinking, culture, and scientific reasoning.

Team Assignment

Examine the core curriculum at your institution and decide how it suggests that your school defines a university education. Write an extended definition that could be used on your school's Web site on materials that your school sends to potential students that help identify your school's philosophy and goals for its students.

Assignments

Part 1: Short Assignments

The following short assignments can be completed either orally or in writing. Unless a team project is specifically indicated, an assignment can be either a team or an individual effort. Your instructor will give you specific directions.

1. Definition

Using the guidelines in this chapter, discuss the relative effectiveness of the following short definitions. Speculate on the likely audience the definitions are addressing.

a. **Afforestation**—the process of establishing trees on land that has lacked forest cover for a very long period of time or has never been forested

b. **Carbon cycle**—the term used to describe the flow of carbon (in various forms, such as carbon dioxide [CO_2], organic matter, and carbonates) through the atmosphere, ocean, terrestrial biosphere, and lithosphere

(continued)

c. **Feebates**—systems of progressive vehicle taxes on purchases of less efficient new vehicles and subsidies for more efficient new vehicles

d. **Greenhouse gases**—gases, including water vapor, CO_2, CH_4, nitrous oxide, and halocarbons that trap infrared heat, warming the air near the surface and in the lower levels of the atmosphere

e. **Mitigation**—a human intervention to reduce the sources of or to enhance the sinks of greenhouse gases

f. **Permafrost**—soils or rocks that remain below 0°C for at least two consecutive years

g. **Temperate zones**—regions of the Earth's surface located above 30° latitude and below 66.5° latitude

h. **Wet climates**—climates where the ratio of mean annual precipitation to potential evapotranspiration is greater than 1.0

Adapted from U.S. Climate Change Science Program. (November 2007.) The First State of the Carbon Cycle Report (SOCCR): *The North American Carbon Budget and Implications for the Global Carbon Cycle*. Synthesis and Assessment Product 2.2. http://www.climatescience.gov/Library/sap/sap2-2/final-report/sap2-2-final-glossary.pdf/

2. Formal Definition

Create formal sentence definitions of the following terms. Remember to include the class and distinguishing features:

- Automated Teller Machine (ATM)
- Digital Video Disc (DVD)
- Web site
- Job Interview

3. Defining a Concept

Concepts often require expanded definitions. Using the ABC format described in this chapter, write a one-page definition of one of the following concepts:

- Community
- Honesty
- Professionalism
- Respect

4. Description

Write a description of a piece of equipment or furniture located in your classroom or brought to class by your instructor—for example, a classroom chair, an overhead projector, a screen, a three-hole punch, a mechanical pencil, or a computer mouse. Write the description for a reader totally unfamiliar with the item.

5. Description

Write a description of a piece of equipment that would be used in a hobby or activity in which you regularly take part. Write the description for someone who has just taken up the activity.

Part 2: Longer Assignments

These assignments test your ability to write the two patterns covered in this chapter—definitions and descriptions. Specifically, follow these guidelines:

- Write each exercise in the form of a letter report or memo report, as specified.
- Follow organization and design guidelines given in Chapters 1 and 3, especially concerning the ABC format (**A**bstract/**B**ody/**C**onclusion) and the use of headings. Chapter 7 gives rules for short reports, but such detail is not necessary to complete the assignments here.
- Fill out a Planning Form (at the end of the book) for each assignment.

6. Technical Definitions in Your Field

Select a technical area in which you have taken course work or in which you have technical experience. Now assume that you are employed as an outside consulting expert, acting as a resource in your particular area to a Human Resources manager not familiar with your specialty. For example, a food-science expert might provide information related to the dietary needs of oil workers working on an offshore rig for three months; a business or management expert might report on a new management technique; an electronics expert might explain the operation of some new piece of equipment that the organization is considering buying; a computer programmer might explain some new piece of hardware that could provide supporting services to the client; and a legal expert might define *sexism in the workplace* for the benefit of the client's human resources professionals.

For the purpose of this report, develop a context in which you would have to define terms for an uninformed reader. Incorporate one expanded definition and at least one sentence definition into your report.

7. Description of Equipment in Your Field

Select a common piece of laboratory, office, or field equipment with which you are familiar. Now assume that you must write a short report to your supervisor, who wants it to contain a thorough physical description of the equipment. Later, he or she plans to incorporate your description into a training manual for those who must understand how to use, and perform minor repairs on, the equipment. For the body of your description, choose either a part-by-part physical description or a thorough description of functions.

8. Description of Position in Your Field

Interview a friend or colleague about the specific job that person holds. Make certain it is a job that you yourself have not had. On the basis of data collected in the interview, write a thorough description of the person's position—including major responsibilities, reporting relationships, educational preparation, and experience required.

⑦ 9. Ethics Assignment

Although definitions and descriptions may appear neutral, they may be used to promote a point of view or to advance an argument on a controversial issue. Examine the following definitions of *global warming* from various sources on the Internet, and find and read each organization's home page. Can you see implied biases in the definition, or does the definition appear neutral? Does this bias or neutrality support the general goals of the organization that published the definition?

In a short essay, compare the definitions and identify the source of each one as well as any apparent bias in the original source. Discuss whether the definitions have been written to support their sources' points of view.

US Geological Service
National Wetlands Research Center
Global Warming—An increase of the earth's temperature by a few degrees resulting in an increase in the volume of water which contributes to sea-level rise.

"The Fragile Fringe: Glossary" <http://www.nwrc.usgs.gov/fringe/glossary.html>. Oct. 4, 2007.

Climate Change Central
Global Warming—Strictly speaking, global warming and global cooling refer to the natural warming and cooling trends that the earth has experienced all through its history. However, the term "global warming" has become a popular term encompassing all aspects of the global warming problem, including the potential climate changes that will be brought about by an increase in global temperatures.

"Glossary of Terms." Retrieved November 2007 from <http://www.climatechangecentral.com/default.asp?V_DOC_ID=849>, n.d.

Minnesota Pollution Control Agency
Global Warming—An increase in the Earth's temperature caused by human activities, such as burning coal, oil and natural gas. This releases carbon dioxide, methane, and other greenhouse gases into the atmosphere. Greenhouse gases form a blanket around the Earth, trapping heat and raising temperatures on the ground. This is steadily changing our climate.

"MPCA Glossary" <http://www.pca.state.mn.us/gloss/glossary.cfm?alpha=G&header=1&glossaryCat=0>, n.d.

Washington Council on International Trade
Global Warming—Heating that occurs when carbon dioxide traps the Sun's heat near Earth's surface, causing Earth's temperature to rise.

"Trade is" <http://www.wcit.org/tradeis/glossary.htm>, n.d.

---

10. International Communication Assignment

In the global marketplace, companies are using illustrations and images to avoid expensive translation. Find examples of descriptions that use illustrations extensively. If possible, find descriptions in multiple languages, such as those in owner's manuals. (Focus on the descriptions of objects, not on instructions.) Analyze the illustrations for their effectiveness as descriptions. How important is text to the illustrations? Could the illustrations serve as descriptions without the text? If you have a document that is in multiple languages, do the illustrations differ from one version to the next? Write an essay that discusses the relationship of text and illustrations in descriptions. Include a discussion of whether you think companies should try to make their descriptions text-free.

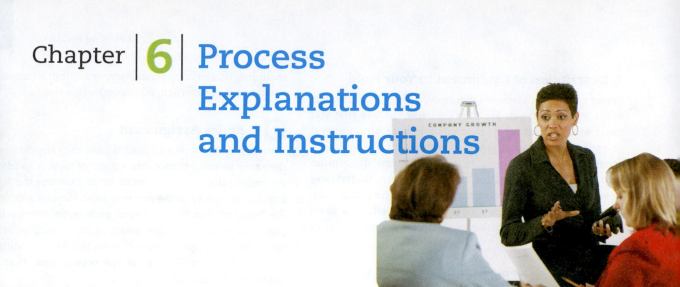

Chapter | 6 | Process Explanations and Instructions

>>> Chapter Outline

Instructions and process explanations share an important common bond. Both must accurately describe a series of steps leading toward a specific result. Yet they differ in purpose, audience, and format. This chapter explores these similarities and differences and gives specific guidelines for developing both types.

>>> Process Explanations versus Instructions

Use a process explanation like the one in Figure 6–1 to help readers *understand* what has been, is being, or will be done. Instructions like the ones in Figure 6–2 show readers how to perform the *process* themselves.

Process explanations are appropriate when the reader must be informed about the action but does not need to perform it. If you suspect a reader may in fact be a *user* (i.e., someone who uses your document to perform the process), always write instructions. Figure 6–3 provides a list of contrasting features of process explanations and instructions.

Process explanations provide information for interested readers who do not need instructional details. Although explaining a process sometimes may be the sole purpose of your document, as in Figure 6-1, often you use process explanation as a pattern of organization within a document with a larger purpose. In both cases, you are writing for a reader who wants to know what has happened or will happen but does not need to perform the process.

Instructions provide users with a road map to *do* the procedure, not just understand it—that is, someone must complete a task on the basis of words and pictures you provide. Clearly, instructions present you, the writer, with a much greater challenge and risk. The reader must be able to replicate the procedure without error and, most importantly, with full knowledge of any dangers. In each case, your instructions must explain steps so thoroughly that the reader will be able to replicate the process without having to speak in person with the writer of the instructions. The next two sections give rules for preparing both process explanations and sets of instructions.

MEMORANDUM

DATE: November 7, 2011
TO: Leonard Schwartz
FROM: Cathy Vir
SUBJECT: New E-mail System

States purpose clearly.

Yesterday I met with Jane Ansel, the installation manager at BHG Electronics, about our new e-mail system. She explained the process by which the system will be installed. As you requested, this memo summarizes what I learned about the setup process.

BHG technicians will be at our offices on November 18 to complete the following tasks:

Describes five main tasks, using parallel grammatical form.

1. Removing old cable from the building conduits
2. Laying cable to link the remaining unconnected terminals with the central processing unit in the main frame
3. Installing software in the system that gives each terminal the capacity to operate the new e-mail system
4. Testing each terminal to make sure the system can operate from that location
5. Instructing selected managers on the use of the system

Confirms the follow-up activities they have already discussed.

As you and I have agreed, when the installation is complete, I will send a memo to all office employees who will have access to e-mail. That memo will discuss setup procedures that each employee must complete before they are able to use their new e-mail accounts.

Gives reader opportunity to respond.

Please let me know if you have further suggestions about how I can help make our transition to the new e-mail system as smooth as possible.

■ **Figure 6–1** ■ Process explanation: electronic mail

MEMORANDUM

DATE: November 20, 2011
TO: All Employees with Access to New E-mail System
FROM: Cathy Vir
SUBJECT: Instructions for Setting Up New E-mail Account

Earlier this month, we had a new e-mail system installed that will be used begin-ning December 1, 2011. This memo provides instructions on how to set up your new e-mail account and how to migrate all of your archived e-mail so that it will be ready for use when the new system goes into effect.   *Gives clear purpose.*

Please follow the step-by-step instructions below for proper setup of your e-mail and migration of your saved e-mail to the new system.   *Identifies result of steps.*

1. Double-click the **E-mail** icon.
2. Use the same **Username** and **Password** that you have used most re-cently with the old e-mail system.   *Limits each step to one action.*
3. Select the **Accounts** menu.
4. Select the **Account Option**s sub-menu.

RESULT: A window will open that prompts an **Account Name** and **Account Type.**   *Separates results from actions.*

5. Enter a name (i.e., "Mail").
6. Use the drop-down menu to select **IMAP4** as the Account Type.
7. Click *Next.*

RESULT: You will be prompted to enter an **Incoming** and **Outgoing Mail Server**.

8. Enter as follows:
    Incoming: www.imap.mglobal.com
    Outgoing: www.smtp.mglobal.com
9. Click *Next.*

RESULT: You will be asked for your **e-mail address**.

10. Use: *yourlastname*@mglobal.com
11. Click *Next.*
12. Click the radio button that reads: **Connect through my Local Area Net-work (LAN)**.
13. Click *Next.*
14. Name your "New Folder" (i.e., "Old Mail")
15. Click the **Finish** button.

Your new account access should now be available, and your old e-mails will move to the new folder that you just named.   *Results if instruc-tions have been followed correctly.*

If you encounter any problems while performing the steps listed above, please contact a member of our IT staff for assistance.   *Shows reader how to get more information.*

■ **Figure 6–2** ■ Instructions: electronic mail

PROCESS EXPLANATIONS

Purpose: Explain a sequence of steps in such a way that the reader understands a process

Format: Use paragraph descriptions, listed steps, or some combination of the two

Style: Use *objective* point of view ("2. The operator started the engine …"), as opposed to *command* point of view ("2. Start the engine …")

INSTRUCTIONS

Purpose: Describe a sequence of steps in such a way that the reader can *perform* the sequence of steps

Format: Employ numbered or bulleted lists, organized into subgroups of easily understandable units of information

Style: Use *command* point of view ("3. Plug the phone jack into the recorder unit"), as opposed to *objective* point of view ("3. The phone jack was plugged into the recorder unit")

>>> Guidelines for Process Explanations

You have already learned that process explanations are aimed at persons who must understand the process, not perform it. Process explanations often have the following purposes:

- Describing an experiment
- Explaining how a machine works
- Recording steps in developing a new product
- Describing procedures to ensure compliance with regulations
- Describing what will happen during a medical procedure

>> Process Guideline 1: Know Your Purpose and Your Audience

Your intended purpose and expected audience influence every detail of your explanation. Following are some preliminary questions to answer before writing:

- Are you supposed to give just an overview, or are details needed?
- Do readers understand the technical subject, or are they laypersons?
- Do readers have mixed technical backgrounds?
- Does the process explanation supply supporting information (perhaps in an appendix), or is it the main part of the document?

When process explanations are directed to a mixed technical audience, write for your least technical readers.

>> Process Guideline 2: Follow the ABC Format

In Chapter 1, you learned about the ABC format (**A**bstract/**B**ody/**C**onclusion) that applies to all documents. The abstract gives a summary, the body supplies details, and the conclusion provides a wrap-up or leads to the next step in the communication process. Whether a process explanation forms all or part of a document, it usually subscribes to the following version of the three-part ABC plan:

- The **Abstract** component includes three background items:

  1. Purpose statement
  2. Overview or list of the main steps that follow
  3. List of equipment or materials used in the process

- The **Body** component of the process explanation moves logically through the steps of the process. By definition, all process explanations follow a *chronological,* or step-by-step, pattern of organization. These steps can be conveyed in two ways, paragraphs or a list of steps.

- The **Conclusion** component of a process explanation keeps the process from ending abruptly with the last step. Here you should help the reader put the steps together into a coherent whole. In Figure 6–1 (p. 110), the last two paragraphs identify the process outlined in the memo as part of a larger change within the organization. When the process explanation is part of a larger document, you can show how the process fits into a larger context.

>> Process Guideline 3: Use an Objective Point of View

Process explanations describe a process rather than direct how it is to be done. Thus they are written from an objective point of view—not from the personal *you* or *command* point of view common to instructions. Note the difference in these examples:

> *Process:* The concrete is poured into the two-by-four frame.
>
> *or*
>
> The technician pours the concrete into the two-by-four frame.
>
> *Instructions:* Pour the concrete into the two-by-four frame.

The process excerpts *explain* the steps, whereas the instructions excerpt *gives a command* for completing the instructions.

>> Process Guideline 4: Choose the Right Amount of Detail

Only a thorough audience analysis will tell you how much detail to include. Figure 6–1 (p. 110), for example, could contain much more detail about the substeps for installing the e-mail system, but the writer decided that the recipient would not need more technical detail.

>> Process Guideline 5: Use Flowcharts for Complex Processes

Some process explanations contain steps that are occurring at the same time. In this case, you may want to supplement a paragraph or list explanation with a flowchart. Such charts

■ **Figure 6–4** ■
Flowchart from explanation of site investigation process

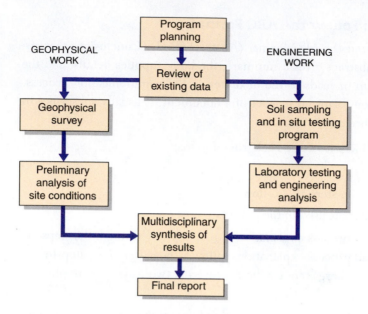

mytechcommlab

For an example of an explanation of procedures see Procedures 2: Pest Control.

use boxes, circles, and other geometric shapes to show progression and relationships among various steps. Figure 6–4 is a flowchart that clarifies how multiple steps in a process will be occurring at the same time.

>>> Guidelines for Instructions

A survey of technical communication managers found that instructional materials, such as manuals and online help remain the most common and most important documents in the field.[1] Thus most writers would benefit from being able to create clear instructions.

Both process explanations and instructions are organized by time, but the similarity stops there. Instructions walk readers through the process so that they can do it, not just understand it. It is one thing to explain the process by which a word processing program works; it is quite another to write a set of instructions for using that word processing program. This section explores the challenge of writing instructions by giving you some basic writing and design guidelines.

These guidelines for instructions also apply to complete operating *manuals,* a document type that many technical professionals will help to write during their careers. Those manuals include the instructions themselves, as well as related information, such as (1) features, (2) physical parts, and (3) troubleshooting tips.

mytechcommlab

See Instructions 1 for an example written for multiple audiences.

>> Instructions Guideline 1: Select the Correct Technical Level

Know exactly who will read your instructions. Are readers technicians, engineers, managers, general users, or some combination of these groups? Once you answer this question, select language that every reader can understand. If, for example, the instructions

---

[1]Kenneth T. Rainey, Roy K. Turner, and David Dayton, "Do Curricula Correspond to Managerial Expectations? Core Competencies for Technical Communicators," *Technical Communication* 52 (2005): 321–352.

include technical terms or names of objects that may not be understood, use the techniques of definition and description discussed in Chapter 5.

>> Instructions Guideline 2: Provide Introductory Information

Like process explanations, instructions follow the ABC format (**A**bstract/**B**ody/**C**onclusion) described in Chapter 1. The introductory (or abstract) information should include (1) a purpose statement, (2) a summary of the main steps, and (3) a list or an illustration giving the equipment or materials needed (or a reference to an attachment with this information). These three items set the scene for the procedure itself.

Besides these three "musts," you should consider whether some additional items might help set the scene for your user:

- Pointers that help with installation
- Definitions of terms
- Theory of how something works
- Notes, cautions, warnings, or dangers that apply to all steps

>> Instructions Guideline 3: Use Numbered Lists in the Body

A simple format is crucial to the body of the instructions—that is, the steps themselves. Most users constantly go back and forth between these steps and the project to which they apply. Thus, you should avoid paragraph format and instead use a simple numbering system.

>> Instructions Guideline 4: Group Steps under Task Headings

Readers prefer that you group together related steps under headings, rather than present an uninterrupted "laundry list" of steps. Figure 6–5 shows how this technique has been used in one section of a fairly long set of instructions for operating a scanner. Given the number of steps in this case, the writer has used a separate numbering system within each grouping.

>> Instructions Guideline 5: Place One Action in a Step

A common error is to bury several actions in a single step. This approach can confuse and irritate readers. Instead, break up complex steps into discrete units.

>> Instructions Guideline 6: Lead Off Each Action Step with a Verb

Instructions should include the *command* form of a verb at the start of each step. This style best conveys a sense of action to your readers. Figure 6–2 on page 111 uses command verbs consistently for all steps throughout the procedures.

■ **Figure 6–5** ■

Grouped steps for
instructions from
installing and
operating a
scanner

1. **Turning on Your Scanner**
   a. Locate the On/Off switch on the front of the scanner.
   b. Switch to the On position.

2. **Scanning**
   a. Open the scanning program by double-clicking the desktop icon.
   b. Place a piece of paper on the scanner bed, in the upper right hand corner.
   c. Select *Scan Document.*
   d. Click *Preview Document.*

   NOTE: This will take 15–20 seconds.

   e. Click and drag the edges of the crop box to fit the document.
   f. Click *Scan.*

   RESULT: The scanner will scan the selected area of the preview image.

3. **Saving Scanned Files**
   a. To save your scanned file(s), go to File > Save.
   b. Enter a name for your document.
   c. Choose to save it as either a JPG (for pictures) or PDF (for text).
   d. Find the file that you want to save the document in.
   e. Click *Save.*

>> **Instructions Guideline 7:** Remove Extra Information from the Step

Sometimes you may want to follow the command sentence with an explanatory sentence or two. In this case, distinguish such helpful information from actions by giving it a label, such as *Note* or *Result* as in Figure 6–2.

>> **Instructions Guideline 8:** Use Bullets or Letters for Emphasis

Sometimes you may need to highlight information, especially within a particular step. Avoid using numbers for this purpose, because you are already using them to signify steps. Bullets work best if there are just a few items; letters are best if there are many, especially if they are in a sequence. In particular, consider using bullets at any point at which users have an *option* as to how to respond.

>> **Instructions Guideline 9:** Emphasize Cautions, Warnings, and Dangers

Instructions often require alerts that draw attention to risks in using products and equipment. Your most important obligation is to highlight such information. Because professional associations and individual companies differ in the way they define terms associated with risk, make sure the alerts in your document follow the appropriate guidelines for your readers. If you have no specific guidelines, however, the following definitions can serve as "red flags" to the reader. The level of risk increases as you move from 1 to 3:

mytechcommlab

Warning boxes are used effectively in Operating Instructions 1.

1. **Caution:** possibility of damage to equipment or materials
2. **Warning:** possibility of injury to people
3. **Danger:** probability of injury or death to people

If you are not certain that these distinctions will be understood by your readers, define the terms *caution, warning,* and *danger* in a prominent place before you begin your instructions.

As for placement of the actual cautions, warnings, or danger messages, your options are as follows:

- **Option 1:** *In a separate section, right before the instructions begin.* This approach is most appropriate when you have a list of general warnings that apply to much of the procedure or when one special warning should be heeded throughout the instructions—for example: "WARNING: Keep main breaker on *off* during entire installation procedure."

- **Option 2:** *In the text of the instructions.* This approach works best if the caution, warning, or danger message applies to the step that immediately follows it. Thus, users are warned about a problem *before* they read the step to which it applies.

- **Option 3:** *Repeatedly throughout the instructions.* This strategy is preferable with instructions that repeatedly pose risk to the user. For example, steps 4, 9, 12A, and 22—appearing on several different pages—may *all* include the hazard of fatal electrical shock. Your *danger* notice should appear in each step, as well as in the introduction to the document.

Give information about potential risks *before* the operator has the chance to make the mistake. Also, the caution, warning, or danger message can be made visually prominent by using font choices such as underlining, bold, or full caps, or graphic elements such as boxes. Color graphics can be another effective indicator of risk. You have probably seen examples, such as a red flame in a box for fire, a jagged line in a triangle for electrical shock, or an actual drawing of a risky behavior with an X through it.

The International Organization for Standardization (ISO) established international standards for safety alerts in ISO 3864, and the American National Standards Institute (ANSI) established domestic standards for safety alerts in ANSI Z535. If the organization you work for complies with ISO or ANSI, you should make sure that you are using the most recent version of the appropriate standards to reinforce the message in your text about cautions, warnings, and dangers.

>> Instructions Guideline 10: Keep a Simple Style

Perhaps more than any other type of technical communication, instructions must be easy to read. Readers expect a no-nonsense approach to writing that gives them required information without fanfare. Following are some useful techniques:

- Keep sentences short, with an average length of fewer than 10 words.
- Use informal definitions (often in parentheses, like this) to define terms not understood by all readers.

- Never use a long word when a short one will do.
- Be specific and avoid words with multiple interpretations (*frequently, seldom, occasionally,* etc.).

>> Instructions Guideline 11: Use Graphics

Illustrations are essential for instructions that involve equipment. Place an illustration next to every major step when (1) the instructions or equipment is quite complicated or (2) the audience may contain unskilled readers or people who are in a hurry. Such word–picture associations create a page design that is easy to follow.

Another useful graphic in instructions is the table. Sometimes within a step you must show correspondence between related data. For example, the instructions below include a spot table.

Step 1: Use pyrometric cones to determine when a kiln has reached the proper temperature for firing pottery. Common cone ratings are as follows:

Cone 018	1200°F
Cone 07	1814°F
Cone 06	1859°F
Cone 04	1940°F

>> Instructions Guideline 12: Test Your Instructions for Usability

Testing instructions for *usability* ensures that your users are able to follow them easily.[2] When you design for usability, you should be focused primarily on the user, not the product itself. This is true whether you are designing a document, software, a computer interface, or a piece of machinery. Products that are usable have the following qualities:

- Learning them is easy.
- Operating them requires the minimum number of steps.
- Remembering how to use them is easy.
- Using them satisfies the user's goals.

Usability does not happen automatically, but should be a concern from the earliest stages of the design of products and documentation.

Professional writers often test their instructions on potential users before completing the final draft. You can adapt the following user-based approach to testing assignments in this class and projects in your career. Specifically, follow these four steps:

1. Team up with another class member (or a colleague on the job). This person should be unfamiliar with the process and should approximate the technical level of your intended audience.

---

[2]Adapted from Carol M. Barnum, *Usability Testing and Research* (New York: Longman, 2002).

2. Give this person a draft of your instructions and provide any equipment or materials necessary to complete the process. For the purposes of a class assignment, this approach works only for a simple process with little equipment or few materials.

3. Observe your colleague following the instructions you provide. You should record both your observations and any verbal responses this person makes while moving through the steps.

4. Revise your instructions to solve problems your user encountered during the test.

>>> Chapter Summary

Both process explanations and instructions share the same organization principle: time. That is, both relate a step-by-step description of events. Process explanations address an audience that wants to be informed but does not need to perform the process itself. Instructions are geared specifically for persons who must complete the procedure themselves.

In writing good process explanations, follow these basic guidelines:

1. Know your purpose and audience.
2. Follow the ABC format.
3. Use an objective point of view.
4. Choose the right amount of detail.
5. Use flowcharts for complex processes.

For instructions, follow these 12 rules:

1. Select the correct technical level.
2. Provide introductory information.
3. Use numbered lists in the body.
4. Group similar steps under heads.
5. Place one action in a step.
6. Lead off each action step with a verb.
7. Remove extra information from the step.
8. Use bullets or letters for emphasis.
9. Emphasize cautions, warnings, and dangers.
10. Keep a simple style.
11. Use graphics.
12. Test your instructions.

>>> Learning Portfolio

Collaboration at Work   A Simple Test for Instructions

General Instructions

Each Collaboration at Work exercise applies strategies for working in teams to chapter topics. The exercise assumes you (1) have been divided into teams of about three to six students, (2) will use team time inside or outside of class to complete the case, and (3) will produce an oral or written response. For guidelines about writing in teams, refer to Chapter 2.

Background for Assignment

Writing instructions presents a challenge. The main problem is this: Although writers may have a good understanding of the procedure for which they are designing instructions, they have trouble adopting the perspective of a reader unfamiliar with the procedure. One way to test the effectiveness of instructions is to conduct your own usability test. The following exercise determines the clarity of instructions written by your team by asking another team to follow the instructions successfully.

Team Assignment

In this exercise, your team prepares a list of instructions for drawing a simple figure or object. The purpose is to write the list so clearly and completely that a classmate could draw the figure or object without knowing its identity. Following are instructions for completing the assignment:

1. Work with your team to choose a simple figure or object that requires only a relatively short set of instructions to draw. (Note: Use a maximum of 15 steps.)
2. Devise a list of instructions that your team believes cannot be misunderstood.
3. Test the instructions within your own team.
4. Exchange instructions with another team.
5. Attempt to draw the object for which the other team has written instructions. (Note: Perform this test without knowing the identity of the object.)
6. Talk with the other team about problems and suggestions related to the instructions.
7. Discuss general problems and suggestions with the entire class.

Assignments

Part 1: Short Assignments

These assignments can be completed either as individual exercises or as team projects, depending on the instructions you are given in class.

1. Writing a Process Explanation

Your college or university has decided to evaluate the process by which students are advised about and registered for classes. As part of this evaluation, the registrar has asked a select team of students—you among them—to explain the actual process each of you went through individually during the last advising/registration cycle. These case studies collected from individual students—the customers—will be transmitted directly to a collegewide committee studying registration and advising problems.

Your job is to give a detailed account of the process. Remain as objective as possible without giving opinions. If you had problems during the process, the facts you relate will speak for themselves. Simply describe the process you

personally experienced, and then let the committee members judge for themselves whether the steps you describe should be a part of the process.

2. Writing Instructions

In either outline or final written form, provide a set of instructions for completing assignments in this class. Consider your audience to be another student who has been ill and missed much of the term. You have agreed to provide her with an overview that will help her to plan and then write any papers she has missed.

Your instructions may include (1) highlights of the writing process from Chapter 1 and (2) other assignment guidelines provided by your instructor in the syllabus or in class. Remember to present a generic procedure for all assignments in the class, not specific instructions for a particular assignment.

> **mytechcommlab**
> Analyze User Guide 2 for ABC format and good visual design.

Part 2: Longer Assignments

These assignments test your ability to write and evaluate the two patterns covered in this chapter—process explanations and instructions. Specifically, follow these guidelines:

- Write each exercise in the form of a letter report or memo report, as specified.
- Follow organization and design guidelines given in Chapters 1 and 3, especially concerning the ABC format (**A**bstract/**B**ody/**C**onclusion) and the use of headings.
- Fill out a Planning Form (at the end of the book) for each assignment.

3. Evaluating a Process Explanation

Using a textbook in a technical subject area, find an explanation of a process—for example, a physics text might explain the process of waves developing and then breaking at a beach, an anatomy text might explain the process of blood circulating, or a criminal justice text might explain the process of processing fingerprints.

Keeping in mind the author's purpose and audience, evaluate the effectiveness of the process explanation as presented in the textbook. Submit your evaluation in the form of a memo report to your instructor in this writing course, along with a copy of the textbook explanation.

For the purposes of this assignment, assume that your writing instructor has been asked by the publisher of the text you have chosen to review the book as an example of good or bad technical writing. Thus, your instructor would incorporate comments from your memo report into his or her comprehensive evaluation.

4. Writing a Process Explanation

Conduct a brief research project in your campus library. Specifically, use company directories, annual reports, or other library sources to find information about a company or other organization that could hire students from your college.

In a memo report to your instructor, (1) explain the process you followed in conducting the search and (2) provide an outline or paragraph summary of the information you found concerning the company or organization. Assume that your report will become part of a volume your college is assembling for juniors and seniors who are beginning their job search. These students will benefit both from information about the specific organization you chose and from an explanation of the process that you followed in getting the information, because they may want to conduct research on other companies.

5. Evaluation of Instructions

Find a set of operating or assembly instructions for a DVD player, microwave oven, CD player, computer, timing light, or other electronic device. Evaluate all or part of the document according to the criteria for instructions in this chapter.

Assume that you are the manager of the technical publications department of the company that produced the electronic device. Write a memo report on your findings and send it, along with a copy of the instructions, to Natalie Bern. As a technical writer in your department, Natalie wrote the set of instructions. In your position as Natalie's supervisor, you are responsible for evaluating her work. Use your memo report either to compliment her on the instructions or to suggest modifications.

6. User Test of Instructions

Find a relatively simple set of instructions. Then ask another person to follow the instructions from beginning to end. Observe the person's activity, keeping notes on any problems she or he encounters.

Use your notes to summarize the effectiveness of the instructions. Present your summary as a memo report to Natalie Bern, using the same situational context as described in Assignment 5—that is, as Natalie's boss, you are to give her your evaluation of her efforts to produce the set of instructions.

7. Writing Simple Instructions

Choose a simple office procedure of 20 or fewer steps (e.g., changing a printer cartridge, cleaning a computer mouse, adding dry ink to a copy machine, adding paper to a laser printer). Then write a simple set of instructions for this process in the form of a memo report. Your readers are assistants at the many offices of a large national firm. Consider them to be new employees who have no background or experience in office work and no education beyond high school. You are responsible for their training.

8. Writing Complex Instructions, with Graphics—Team Project

Complete this assignment as a team project (see the guidelines for team work in Chapter 2). Choose a process connected with college life or courses—for example, completing a lab experiment, doing a field test, designing a model, writing a research paper, getting a parking sticker, paying fees, or registering for classes.

Using memo report format, write a set of instructions for students who

mytechcommlab

For more practice, do the Revising Installation Instructions Activity.

have never performed this task. Follow all the guidelines in this chapter. Include at least one illustration (along with warnings or cautions, if appropriate). If possible, conduct a user test before completing the final.

⑦ 9. Ethics Assignment

Examine a set of instructions for a household or recreational device that—either in assembly or use—poses serious risk of injury or death. Evaluate the degree to which the manufacturer has fulfilled its ethical responsibility to inform the user of such risk. You may want to consider the following questions:

a. Are risks adequately presented in text and/or graphic form?

b. Are risk notices appropriately placed in the document?

c. Is the document designed such that a user reading quickly could locate cautions, warnings, or dangers easily?

If you have highlighted any ethical problems, also suggest solutions to these problems.

10. International Communication Assignment

Sets of instructions may reflect *cultural bias* of a particular culture or country. Such bias may be acceptable if the audience for the instructions shares the same background.

However, cultural bias presents a problem when (1) the audience represents diverse cultures and backgrounds or (2) the instructions must be translated into another language by someone not familiar with cultural cues in the instructions. Following are just a few categories of information that can present cultural bias and possibly cause confusion:

- Date formats
- Time zones
- Types of monetary currency
- Units of measurement
- Address and telephone formats
- Historical events
- Geographic references
- Popular culture references
- Acronyms
- Legal information
- Common objects in the home or office

[Adapted from a list on pp. 129–130 of Nancy L. Hoft's *International Technical Communication: How to Export Information About High Technology* (New York: John Wiley, 1995).]

Choose a set of instructions that reflects several types of cultural bias, such as those included on the previous list. Point out the examples of bias and explain why they might present problems to readers outside a particular culture.

Chapter | 7 | Reports

>>> Chapter Outline

Reports communicate the results of research and activities in an organization. They may also be archived to provide a record of what an organization has accomplished. This chapter provides general guidelines for writing reports, and specific guidelines for the two report formats—informal reports and formal reports, which differ primarily in their length and scope. This chapter also introduces you to four common types of reports.

>>> General Guidelines for Reports

You will write many reports in your career. Reports may be as simple as a single-page form, like an activity report that you submit daily or weekly so that clients can be billed for your time. Or they may be over 100-page long, like an analysis of a major equipment failure. This section provides general guidelines for effective reports.

>> Report Guideline 1: Plan Well Before You Write

As with all types of workplace writing, planning is the most important part of the writing process. For each project, complete the Planning Form at the end of this book to record specific information about these points:

- The document's purpose
- The variety of readers who will receive the document
- The needs and expectations of readers, particularly decision makers
- An outline of the main points to be covered in the body
- Strategies for writing an effective document

>> Report Guideline 2: Separate Fact from Opinion

Avoid any confusion about what constitutes fact or opinion. The safest approach in the report discussion is to move logically from findings to your conclusions and, finally, to your recommendations. Because these terms are often confused, some working definitions are as follows:

- **Findings:** Facts you uncover (e.g., you observed severe cracks in the foundations of two adjacent homes in a subdivision).
- **Conclusions:** Ideas or beliefs you develop based on your findings (e.g., you conclude that foundation cracks occurred because the two homes were built on soft fill, where original soil had been replaced by construction scraps).
- **Recommendations:** Suggestions or action items based on your conclusions (e.g., you recommend that the foundation slab be supported by adding concrete posts beneath it). Though consisting of opinions, recommendations should be grounded in facts presented in the report.

>> Report Guideline 3: Make Text Visually Appealing

Following are three visual devices that help get attention, maintain interest, and highlight important information:

- Bulleted points for short lists (like this one)
- Numbered points for lists that are longer or that include a list of ordered steps
- Frequent use of headings and subheadings

>> Report Guideline 4: Use Illustrations for Clarification and Persuasion

A simple table or figure can sometimes be just the right complement to a technical discussion in the text. Incorporate illustrations into the report body to make technical information accessible and easier to digest.

>> Report Guideline 5: Edit Carefully

The Handbook at the end of this text gives information about editing. For now, remember the following basic guidelines:

- Keep most sentences short and simple.
- Proofread several times for mechanical errors, such as misspellings (particularly personal names).
- Triple-check all cost figures for accuracy.
- Make sure all attachments are included, are mentioned in the text
- Check the format and wording of all headings and subheadings.
- Ask a colleague to check over the report.

Remember—both supervisors and clients will judge you as much on communication skills as they do on technical ability.

>>> Guidelines for Informal Reports

A working definition of informal reports follows:

Informal report: A document that contains about two to five pages of text, not including attachments. It has more substance than a simple letter or memo, but is presented in letter or memo format. It can be directed to readers either outside or inside your organization. If outside, it may be called a *letter report*; if inside, it may be called a *memo report*. In either case, its purpose can be *informative* (to clarify or explain), *persuasive* (to convince), or both.

Following are six guidelines that focus on informal report format:

>> Informal Report Guideline 1: Use Letter or Memo Format

Figure 7–1 shows that letter reports follow about the same format as typical business letters (see Chapter 4). Yet, the format of letter reports differs from that of letters in the following respects:

- The greeting is sometimes left out or replaced by an attention line, especially when your letter report will go to many readers in an organization.
- A report title often comes immediately after the inside address. It identifies the specific project covered in the report.
- Spacing between lines might be single, one-and-one-half, or double, depending on the reader's preference.

Memo reports follow the same format as typical business memos. Both memos and memo reports have a subject line that should engage interest, give readers their first quick look at your topic, and be both specific and concise—for example, "Fracture Problems with Molds 43-D and 42-G" is preferable to "Problems with Molds." Because memo reports are usually longer than memos, they tend to contain more headings than routine memos.

>> Informal Report Guideline 2: Use the ABC Format for Organization

Most technical documents, including informal reports, follow what this book calls the *ABC format*. This approach to organization includes three parts: (1) **A**bstract, (2) **B**ody, and (3) **C**onclusion. The next four guidelines give details on the ABC format as applied to informal reports.

ABC Format: Organization

- **ABSTRACT:** Start with a capsule version of the information most needed by decision makers.
- **BODY:** Give details in the body of the report, where technical readers are most likely to linger a while to examine supporting evidence.
- **CONCLUSION:** Reserve the end of the report for a description or list of findings, conclusions, or recommendations.

>> Informal Report Guideline 3: Create the Abstract as an Introductory Summary

Abstracts should give readers a summary, the "big picture." This text suggests that in informal reports, you label this overview *Introduction, Summary,* or *Introductory Summary,* terms that give the reader a good idea of what the section contains. Just one or two paragraphs in this first section give readers three essential pieces of information:

1. **Purpose** for the report—why are you writing it?
2. **Scope** statement—what range of information does the report contain?
3. **Summary** of essentials—what main information does the reader most want or need to know?

>> Informal Report Guideline 4: Put Important Details in the Body

The body section expands on the outline presented in the introductory summary. If your report goes to a diverse audience, managers often read the quick overview in the

12 Post Street
Houston Texas 77000
(713) 555-9781

April 22, 2010

Big Muddy Oil Company Inc
12 Rankin St
Abilene TX 79224

ATTENTION: Mr. James Smith, Engineering Manager

SHARK PASS STUDY
BLOCK 15, AREA 43-B ◄——— Includes specific title.
GULF OF MEXICO

INTRODUCTORY SUMMARY ◄——— Uses *optional* heading for abstract part of ABC format.

You recently asked our firm to complete a preliminary soils investigation at an offshore rig site. This report presents the tentative results of our study, including major conclusions and recommendations. A longer, formal report will follow at the end of the project.

On the basis of what we have learned so far, it is our opinion that you can safely place an oil platform at the Shark Pass site. To limit the chance of a rig leg punching into the sea floor, however, we suggest you follow the recommendations in this report. ◄——— Draws attention to *main point* of report.

WORK AT THE PROJECT SITE

On April 15 and 16, 2010, BoomCo's engineers and technicians worked at the ◄——— Gives on-site details of project—dates, location, tasks. Block 15 site in the Shark Pass region of the gulf. Using BoomCo's leased drill ship, *Seeker II*, as a base of operations, our crew performed these main tasks:

• Seismic survey of the project study area
• Two soil borings of 40 feet each

Both seismic data and soil samples were brought to our Houston office the next ◄——— Uses *lead-in* to subsections that follow. day for laboratory analysis.

LABORATORY ANALYSIS

On April 17 and 18, our lab staff examined the soil samples, completed bearing capacity tests, and evaluated seismic data. Here are the results of that analysis.

Soil Layers

Our initial evaluation of the soil samples reveals a 7–9 feet layer of weak clay ◄——— Highlights most important point about soil layer—that is the *weak clay*. starting a few feet below the seafloor. Other than that layer, the composition of the soils seems fairly typical of other sites nearby.

M-Global Inc. | 127 Rainbow Lane | Baltimore MD 21202 | 410.555.8175

■ **Figure 7–1** ■ Recommendation report (letter format)

James Smith
April 22, 2010
Page 2

Bearing Capacity

Notes *why* this method was chosen (i.e., reliability).

We used the most reliable procedure available, the XYZ method, to determine the soil's bearing capacity (i.e., its ability to withstand the weight of a loaded oil rig). That method required that we apply the following formula:

Q = $cNv + tY$, where
Q = ultimate bearing capacity
c = average cohesive shear strength
Nv = the dimensionless bearing capacity factor
t = footing displacement
Y = weight of the soil unit

The final bearing capacity figure will be submitted in the final report, after we repeat the tests.

Seafloor Surface

Explains both *how* the mapping procedure was done and *what results* it produced.

By pulling our underwater seismometer back and forth across the project site, we developed a seismic "map" of the seafloor surface. That map seems typical of the flat floor expected in that area of the gulf. The only exception is the presence of what appears to be a small sunken boat. This wreck, however, is not in the immediate area of the proposed platform site.

CONCLUSIONS AND RECOMMENDATIONS

Leads off section with major conclusion, for emphasis.

Restates points (made in body) that support conclusion.

Based on our analysis, we conclude that there is only a slight risk of instability at the site. Although unlikely, it is possible that a rig leg could punch through the sea floor, either during or after loading. We base this opinion on (1) the existence of the weak clay layer, noted earlier and (2) the marginal bearing capacity.

Nevertheless, we believe you can still place your platform if you follow careful rig-loading procedures. Specifically, take these precautions to reduce your risk:

Uses list to emphasize recommendations to *reduce* risk.

1 Load the rig in 10-ton increments, waiting 1 hour between loadings.
2 Allow the rig to stand 24 hours after the loading and before placement of workers on board.
3 Have a soils specialist observe the entire loading process to assist with any emergency decisions if problems arise.

Again mentions tentative nature of information, to prevent misuse of report.

As noted at the outset, these conclusions and recommendations are based on preliminary data and analysis. We will complete our final study in 3 weeks and submit a formal report shortly thereafter.

Maintains contact and shows initiative by offering to *call* client.

BoomCo, Inc., enjoyed working once again for Big Muddy Oil at its Gulf of Mexico lease holdings. I will phone you this week to see if you have any questions about our study. If you need information before then, please give me a call.

Sincerely,

Bartley Hopkins

Bartley Hopkins, Project Manager
BoomCo, Inc.
hg

■ **Figure 7–1** ■ continued

introductory summary and then skip to conclusions and recommendations. Technical readers, however, may look first to the body section(s), where they expect to find supporting details.

- **Use headings generously.**   Each time you change a major or minor point, consider whether a heading change would help the reader.
- **Precede subheadings with a lead-in passage.**   For example, "This section covers these three phases of the field study: clearing the site, collecting samples, and classifying samples." The preceding sentence does for the entire section exactly what the introductory summary does for the entire report—it provides a roadmap for what's ahead.
- **Move from general to specific in paragraphs.**   Start each paragraph with a topic sentence that includes your main point and then give supporting details.

>> Informal Report Guideline 5: Focus Attention in Your Conclusion

Letter and memo reports can end with a section labeled *Findings, Conclusions,* or *Conclusions and Recommendations.* This section gives details about major findings, conclusions, and, if called for, recommendations. People often remember best what they read last, so think hard about what you place at the end of a report.

The precise amount of detail in your conclusion depends on which of these two options you choose for your particular report:

Option 1:   If your major conclusions or recommendations have already been stated in the discussion, then you only need to restate them briefly to reinforce their importance.

Option 2:   If the discussion leads up to, but has not covered, these conclusions or recommendations, then you may want to give more detail in this final section (see Figure 7–1, pp. 126–127).

>> Informal Report Guideline 6: Use Attachments for Less Important Details

The trend today is to avoid lengthy text in informal reports. Replace as much report text as possible with clearly labeled attachments that include items, such as tables, figures, or costs.

>>> Guidelines for Formal Reports

You will write a number of long, formal reports during your career, often with colleagues. This text uses the following working definition for a formal report:

Formal report: A formal report covers complex projects and is directed to readers at different technical levels. Although not defined by length, a formal report usually contains at least 6 to 10 pages of text, not including appendices. It can be directed to readers either inside or outside your organization.

Because formal reports may be longer and more complex than other forms of technical communication, it is important to help your readers navigate through the report. This is accomplished with special front and end materials, clear headings, and other navigation devices like running headers and footers. This section provides guidelines for writing the main parts of a long report and includes a complete long report that follows this chapter's guidelines.

Though formats differ among organizations and disciplines, one approach to good organization applies to all formal reports. This approach is based on three main principles, discussed in detail in Chapter 1:

Principle 1: Write different parts for different readers.

Principle 2: Place important information first.

Principle 3: Repeat key points when necessary.

These apply to formal reports even more than they do to short documents. Because formal reports often have a mixed technical audience, most readers focus on specific sections that interest them most, and few readers have time to wade through a lot of introductory information before reaching the main point.

You can respond to these reader needs by following the ABC format (for Abstract, Body, Conclusion). As noted in Chapter 1, the three main rules are that you should (1) start with an abstract for decision makers, (2) put supporting details in the body, and (3) use the conclusion to produce action. This simple ABC format should be evident in all formal reports, despite their complexity. The particular sections of formal reports fit within the ABC format as shown on the right:

ABC Format: **Formal Report**

- **A**BSTRACT:
  - Cover/title page
  - Letter or memo of transmittal
  - Table of contents
  - List of illustrations
  - Executive summary
  - Introduction
- **B**ODY:
  - Discussion sections
  - [Appendices—appear after text but support Body section]
- **C**ONCLUSION:
  - Conclusions
  - Recommendations

Nine Parts of Formal Reports

The nine parts of the formal report are as follows:

1. Cover/title page
2. Letter or memo of transmittal
3. Table of contents
4. List of illustrations
5. Executive summary
6. Introduction
7. Discussion sections

8. Conclusions and recommendations
9. End material

Following are some guidelines for these parts.

Cover/Title Page

Formal reports are usually bound, often with a cover used for all reports in the writer's organization. Because the cover is the first item seen by the reader, it should be attractive and informative. It usually contains the same four pieces of information mentioned in the following list with regard to the title page.

Inside the cover is the title page, which should include the following four pieces of information:

- Project title (exactly as it appears on the letter/memo of transmittal)
- Your client's name ("Prepared for…")
- Your name and/or the name of your organization ("Prepared by…")
- Date of submission

To make your title page or cover distinctive, you might want to place a simple illustration on it; however, do not clutter the page. Use a visual only if it reinforces a main point and if it can be done simply and tastefully, as in Figure 7–2.

Letter/Memo of Transmittal

Letters or memos of transmittal give the readers a taste of what is ahead. If your formal report is to readers outside your own organization, write a letter of transmittal. If it is to readers inside your organization, write a memo of transmittal.

>> Transmittal Guideline 1: Place the Letter/Memo Immediately after the Title Page

This placement means that the letter/memo is bound with the document, to keep it from becoming separated.

>> Transmittal Guideline 2: Include a Major Point from Report

Readers are heavily influenced by what they read first in reports. Therefore, take advantage of the position of this section by including a major finding, conclusion, or recommendation from the report—besides supplying necessary transmittal information.

>> Transmittal Guideline 3: Acknowledge Those Who Helped You

Recognizing those who have been particularly helpful with your project gives them recognition and identifies you as a team player. It reflects well on you and on your organization.

>> Transmittal Guideline 4: Follow Letter and Memo Conventions

Like other letters and memos, letters and memos of transmittal should be easy to read, inviting readers into the rest of the report. Keep introductory and concluding paragraphs relatively short—no more than three to five lines each. Also, write in a conversational style, free of technical jargon.

Table of Contents

Your contents page acts as an outline. Many readers go there right away to grasp the structure of the report, and then return repeatedly to locate report sections of most interest to them. Guidelines follow for assembling this important component of your report; see p. 137 for an example in Figure 7–2.

>> Table of Contents Guideline 1: Make It Very Readable

The table of contents must be pleasing to the eye so that readers can find sections quickly and see their relationship to each other. Be sure to

- Space items well on the page
- Use indenting to draw attention to subheadings
- Include page numbers for every heading and subheading, unless there are many headings in a relatively short report, in which case you can delete page numbers for all of the lowest level headings listed in the table of contents

>> Table of Contents Guideline 2: Use the Contents Page to Reveal Report Emphases

Be specific yet concise so that each heading listed in the table of contents gives the reader a good indication of what the section contains. Readers associate the importance of report sections with the number of headings and subheadings listed in the table of contents.

>> Table of Contents Guideline 3: Consider Leaving Out Low-Level Headings

In very long reports, you may want to declutter the table of contents by removing lower-level headings. As always, the needs of the readers are the most important criterion to use in making this decision.

>> Table of Contents Guideline 4: List Appendices

Appendices include items, such as tables of data or descriptions of procedures that are inserted at the end of the report. Typically, they are listed at the end of the table of contents. Tabs on the edges of pages help the reader locate these sections.

>> Table of Contents Guideline 5: Use Parallel Form in All Entries

All headings in one section, and sometimes even all headings and subheadings in the report, have parallel grammatical form.

>> Table of Contents Guideline 6: Proofread Carefully

Wrong page numbers and incorrect headings are two common mistakes. Proofread this section carefully.

List of Illustrations

Illustrations within the body of the report are usually listed on a separate page right after the table of contents. Another option is to list them at the bottom of the table of contents page rather than on a separate page. In either case, this list should include the number, title, and page number of every table and figure within the body of the report.

Executive Summary

No formal report would be complete without an executive summary. Consider it a stand-alone section that provides a capsule version of the report and is free of technical jargon. In some cases, a copy of the executive summary may be circulated and filed separate from the report.

>> Executive Summary Guideline 1: Put It on One Page

It is comforting to most readers to know that somewhere within a long report there is one page to which they can turn for an overview. Moreover, a one-page length permits easy distribution at meetings.

Some extremely long formal reports may require that you write an executive summary of several pages or longer. In this case, you can still provide the reader with a section that summarizes the report in less than a page by including a brief *abstract*. The abstract is a condensed version of the executive summary directed to the highest-level decision makers, that is placed right before the executive summary.

>> Executive Summary Guideline 2: Avoid Technical Jargon

Include only that level of technical language the decision makers understand. Do not talk over the heads of the most important readers.

>> Executive Summary Guideline 3: Include Only the Important Conclusions and Recommendations

The executive summary mentions only the major points. An exhaustive list of findings, conclusions, and recommendations can come later at the end of the report.

>> Executive Summary Guideline 4: Avoid References to the Report Body

Avoid the tendency to say that the report provides additional information. It is understood that the executive summary is only a generalized account of the report's contents.

An exception is those instances when you are discussing issues that involve danger or liability in which it may be necessary to add qualifiers in your summary—for example, "As noted in this report, further study will be necessary." Such statements protect you and the client in the event the executive summary is removed from the report and used as a separate stand-alone document.

>> Executive Summary Guideline 5: Use Paragraph Format

Whereas lists are often appropriate for body sections of a report, they can give executive summaries a fragmented effect. Instead, the best summaries create unity with a series of relatively short paragraphs that flow well together. Within a paragraph, there can be a short listing of a few points for emphasis, but the listing should not be the main structural element of the summary.

>> Executive Summary Guideline 6: Write the Executive Summary Last

Only after finishing the report do you have the perspective to write a summary. Sit back and review the report from beginning to end, asking yourself, "What would my readers really need to know if they had only a minute or two to read?" The answer to that question becomes the core of your executive summary.

Introduction

View this section as your chance to prepare both technical and nontechnical readers for the discussion ahead. Give information on the report's purpose, scope, and format, as well as a project description.

>> Introduction Guideline 1: State Your Purpose and Lead into Subsections

The purpose statement for the document should appear immediately after the main introduction heading. Follow it with a sentence that mentions the introduction subdivisions to follow.

>> Introduction Guideline 2: Include a Project Description

Here you must be precise about the project. Depending on the type of project, you may be describing a physical setting, a set of problems that prompted the report study, or some other data. When the project description is too long for the introduction, sometimes it is placed in the body of the report.

>> Introduction Guideline 3: Include Scope Information

This section outlines the precise objectives of the study. Your listing or description should parallel the order of the information presented in the body of the report. Like the project description, this subsection must be accurate in every detail.

>> Introduction Guideline 4: Consider Including Information on Report Format

Often, the scope section lists information as it is presented in the report. If this is not the case, end the introduction with a short subsection on the report format where you can give readers a brief preview of the main sections that follow. In effect, the section acts as a condensed table of contents and may list the report's major sections and appendices.

Discussion Sections

Discussion sections make up the longest part of formal reports and are written for the most technical members of your audience. You can focus on facts and opinions, demonstrating the technical expertise that the reader expects from you.

>> Discussion Guideline 1: Move from Facts to Opinions

As you have learned, the ABC format requires that you start your formal report with a summary of the most important information—that is, you skip to essential conclusions and recommendations. Once into the discussion section, however, you back up and adopt a strategy that parallels the stages of the technical project itself. You begin with data and move toward conclusions and recommendations.

One way to view the discussion is that it should follow the order of a typical technical project, which usually involves the following stages:

First, collect data (e.g., samples, interviews, records).

Second, subject these data to verification or testing (e.g., lab tests, computer analyses).

Third, analyze all the information, using your professional experience and skills to form conclusions (or convictions based on the data).

Fourth, develop recommendations that flow directly from the conclusions you have formed.

Thus, the body of your report gives technical readers the same movement from fact toward opinion that you experience during the project itself.

>> Discussion Guideline 2: Use Frequent Headings and Subheadings

Headings give readers handles by which to grasp the content of your report. Your readers view headings, collectively, as a sort of outline by which they can make their way easily through the report.

>> Discussion Guideline 3: Use Listings to Break Up Long Paragraphs

Long paragraphs full of technical details irritate readers. Use paragraphs for brief explanations, not for descriptions of processes or other details that could be listed.

>> Discussion Guideline 4: Place Excessive Detail in Appendices

Appendices give readers access to supporting information without cluttering the text of the report. Of course, you must refer to appendices in the body of the report and label appendices clearly so that readers can locate them easily.

Conclusions and Recommendations

This section gives readers a description—sometimes in the form of a listing—of all conclusions and recommendations. The points may or may not have been mentioned in the body of the report, depending on the length and complexity of the document. *Conclusions,* on the one hand, are convictions or beliefs based on the findings of your study. *Recommendations,* on the other hand, are actions you are suggesting based on your conclusions. For example, your conclusion may be that there is a dangerous level of toxic chemicals in a town's water supply, and your recommendation may be that the toxic site near the reservoir should be cleaned immediately.

The "Conclusions and Recommendations" section provides a complete list of conclusions and recommendations for technical and management readers, whereas the "Executive Summary" provides a selected list or description of the most important conclusions and recommendations for decision makers. In other words, view the "Conclusions and Recommendations" section as an expanded version of the "Executive Summary." It usually assumes one of these three headings, depending, of course, on the content:

1. Conclusions
2. Recommendations
3. Conclusions and Recommendations

Another option for reports that contain many conclusions and recommendations is to separate this last section into two sections: (1) "Conclusions" and (2) "Recommendations."

End Material

One kind of end material—appendices—is mentioned in the context of the discussion section. Note that formal reports may also contain works-cited pages or bibliographies, which should be included in the end materials. Finally, very long reports may include indices.

>>> Formal Report Example

Figure 7–2 provides a long and formal technical report. The report results from a study completed for the city of Winslow, Georgia. Members of the audience come from both technical and nontechnical backgrounds. Some are full-time professionals hired by the city, whereas others are part-time, unpaid citizens appointed by the mayor to explore

STUDY OF WILDWOOD CREEK

WINSLOW, GEORGIA

Prepared for:

The City of Winslow

Prepared by:

Christopher S. Rice, Hydro/Environmental Engineer
D-Lynn, Inc.

November 28, 2010

Uses graphic on title page to reinforce theme of environmental protection.

■ **Figure 7–2** ■ Formal report

12 Peachtree Street
Atlanta GA 30056
(404) 555-7524

D-Lynn Project #99-119
November 28, 2010

Adopt-a-Stream Program
City of Winslow
300 Lawrence Street
Winslow
Georgia 30000

Attention: Ms. Elaine Sykes, Director

STUDY OF WILDWOOD CREEK
WINSLOW, GEORGIA

Lists project title as it appears on title page.

We have completed our seven-month project on the pollution study of Wildwood Creek. This project was authorized on May 16, 2010. We performed the study in accordance with our original proposal No. 14-P72, dated April 24, 2010.

Gives brief statement of project information.

This report mentions all completed tests and discusses the test results. Wildwood Creek scored well on many of the tests, but we are concerned about several problems—such as the level of phosphates in the stream. The few problems we observed during our study have led us to recommend that several additional tests should be completed.

Provides major point from report.

Thank you for the opportunity to complete this project. We look forward to working with you on further tests for Wildwood Creek and other waterways in Winslow.

Sincerely,

Christopher S. Rice

Christopher S. Rice, P.E.
Hydro/Environmental Engineer

M-Global Inc | 127 Rainbow Lane | Baltimore MD 21202 | 410.555.8175

■ **Figure 7–2** ■ continued

CONTENTS

Uses white space, indenting, and bold to accent organization of report.

■ **Figure 7–2** ■ continued

Includes illustration titles as they appear in text.

ILLUSTRATIONS

FIGURES   PAGE

TABLES

■ **Figure 7–2** ■ continued

2

EXECUTIVE SUMMARY

The City of Winslow hired D-Lynn, Inc., to perform a pollution study of Wildwood Creek. The section of the creek that was studied is a one-mile-long area in Burns Nature Park, from Newell College to U.S. Highway 42. The study lasted seven months.

Summarizes purpose and scope of report.

D-Lynn completed 13 tests on four different test dates. Wildwood scored fairly well on many of the tests, but there were some problem areas—for example, high levels of phosphates were uncovered in the water. The phosphates were derived either from fertilizer or from animal and plant matter and waste. Also uncovered were small amounts of undesirable water organisms that are tolerant to pollutants and can survive in harsh environments.

Describes major findings and conclusions.

D-Lynn recommends that (1) the tests done in this study be conducted two more times, through spring 2011; (2) other environmental tests be conducted, as listed in the conclusions and recommendations section; and (3) a voluntary cleanup of the creek be scheduled. With these steps, we can better analyze the environmental integrity of Wildwood Creek.

Includes main recommendation from report text.

■ **Figure 7–2** ■ continued

3

INTRODUCTION

Gives lead-in to Introduction.

D-Lynn, Inc., has completed a follow-up to a study completed in 2002 by Ware County on the health of Wildwood Creek. This introduction describes the project site, scope of our study, and format for this report.

PROJECT DESCRIPTION

Briefly describes project.

By law, all states must clean up their waterways. The State of Georgia shares this responsibility with its counties. Ware County has certain waterways that are threatened and must be cleaned. Wildwood Creek is one of the more endangered waterways. The portion of the creek that was studied for this report is a one-mile stretch in the Burns Nature Park between Newell College and U.S. Highway 42.

SCOPE OF STUDY

The purpose of this project was to determine whether the health of the creek has changed since the previous study in 2002. Both physical and chemical tests were completed. The nine physical tests were as follows:

Uses bulleted list to emphasize scope of activities.

- Air temperature
- Water temperature
- Water flow
- Water appearance
- Habitat description
- Algae appearance
- Algae location
- Visible litter
- Bug count

The four chemical tests were as follows:

- pH
- Dissolved oxygen (DO)
- Turbidity
- Phosphate

REPORT FORMAT

Provides "map" of main sections in report.

This report includes three main sections:

1. Field Investigation: a complete discussion of all the tests that were performed for the project
2. Test Comparison: charts of the test results and comparisons
3. Conclusions and Recommendations

■ **Figure 7–2** ■ continued

4

FIELD INVESTIGATION

Wildwood Creek has been cited repeatedly for environmental violations in the pollution of its water. Many factors can generate pollution and affect the overall health of the creek. In 2002, the creek was studied in the context of a study of all water systems in Ware County. Wildwood Creek was determined to be one of the more threatened creeks in the county.

The city needed to learn if much has changed in the past nine years, so D-Lynn was hired to perform a variety of tests on the creek. Our effort involved a more in-depth study than that done in 2002. Tests were conducted four times over a seven-month period. The 2002 study lasted only one day.

The field investigation included two categories of tests: physical tests and chemical tests.

PHYSICAL TESTS

The physical tests covered a broad range of environmental features. This section discusses the importance of the tests and some major findings. The Test Comparison section on page 9 includes a table that lists results of the tests and the completion dates. The test types were as follows: air temperature, water temperature, water flow, water appearance, habitat description, algae appearance, algae location, visible litter, and bug count.

← Amplifies information presented later in report.

Air Temperature

The temperature of the air surrounding the creek will affect life in the water. Unusual air temperature for the seasons will determine if life can grow in or out of the water.

Three of the four tests were performed in the warmer months. Only one was completed on a cool day. The difference in temperature from the warmest to coolest day was 10.5°C, an acceptable range.

Water Temperature

The temperature of the water determines which species will be present. Also affected are the feeding, reproduction, and metabolism of these species. If there are one or two weeks of high temperature, the stream is unsuitable for most species. If water temperature changes more than 1° to 2°C in 24 hours, thermal stress and shock can occur, killing much of the life in the creek.

During our study, the temperature of the water averaged 1°C cooler than the temperature of the air. The water temperature did not get above 23°C or below 13°C. These ranges are acceptable by law.

■ **Figure 7–2** ■ continued

5

Water Flow

The flow of the water influences the type of life in the stream. Periods of high flow can cause erosion to occur on the banks and sediment to cover the streambed. Low water flow can decrease the living space and deplete the oxygen supply.

The flow of water was at the correct level for the times of year the tests were done—except for June, which had a high rainfall. With continual rain and sudden flash floods, the creek was almost too dangerous for the study to be performed that month.

In fact, in June we witnessed the aftermath of one flash flood. Figure 1 shows the creek with an average flow of water, and Figure 2 shows the creek during the flood. The water's average depth is 10 inches. During the flash flood, the water level rose and fell 10 feet in about one hour. Much dirt and debris were washing into the creek, while some small fish were left on dry land as the water receded.

Incorporates graphic into page of text.

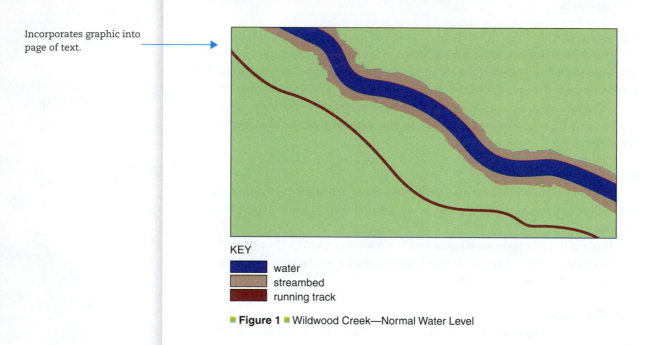

KEY
- water
- streambed
- running track

■ **Figure 1** ■ Wildwood Creek—Normal Water Level

■ **Figure 7–2** ■ continued

6

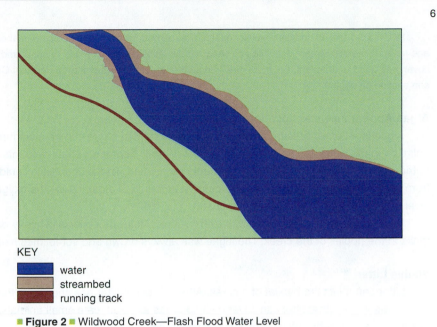

KEY
- ■ water
- ■ streambed
- ■ running track

■ **Figure 2** ■ Wildwood Creek—Flash Flood Water Level

Water Appearance

The color of the water gives a quick but fairly accurate view of the health of the creek. If the water is brown or dirty, then silt or human waste may be present. Black areas of water may contain oil or other chemical products.

On each of the four test days, the water was always clear. Thus the appearance of the creek water was considered excellent.

Habitat Description

The habitat description concerns the appearance of the stream and its surroundings. An important criterion is the number of pools and the number of ripples—that is, points where water flows quickly over a rocky area. Both pools and ripples provide good locations for fish and other stream creatures to live and breed.

In describing habitat, D-Lynn also evaluates the amount of sediment at the bottom of the stream. Too much sediment tends to cover up areas where aquatic life lays eggs and hides them from predators. We also evaluate the stability of the stream banks; a stable bank indicates that erosion has not damaged the habitat. Finally, we observe the amount of stream cover. Such vegetation helps keep soil in place on the banks.

Elaborates on importance of information shown in Table 1. Description parallels five items in table.

■ **Figure 7–2** ■ continued

7

Wildwood Creek tested fairly well for habitat. The number of pools and ripples was about average for such creeks. Stream deposits and stream bank stability were average to good, and stream cover was good to excellent. For more detail about test results, see the chart in the Test Comparison section on page 9.

Algae Appearance and Location

Algae is naturally present in any creek. The amount of algae can be a warning of pollution in the water. If algae is growing out of control, disproportionate amounts of nutrients such as nitrogen or phosphate could be present. These chemicals could come from fertilizer washed into the creek. Excessive amounts of algae cause the oxygen level to drop when they die and decompose.

During the four studies, algae was everywhere, but it was especially heavy on the rocks in the ripples of the creek. The algae was always brown and sometimes hairy.

Visible Litter

Litter can affect the habitat of a creek. Although some litter has chemicals that can pollute the water, other litter can cover nesting areas and suffocate small animals. Whether the litter is harmful or not, it is always an eyesore.

On all four test dates, the litter we saw was heavy and ranged from tires to plastic bags. Some of the same trash that was at the site on the first visit was still there seven months later.

Gives specific details that support the report's conclusions and recommendations, which come later.

Bug Count

The bug count is a procedure that begins by washing dirt and water onto a screen. As water drains, the dirt with organisms is left on the screen. The bugs are removed and classified. Generally, the lower the bug count, the higher the pollution levels. Bug counts were considered low to average.

Two types of aquatic worms were discovered every time during our count, but in relatively small amounts. In addition, the worms we observed are very tolerant of pollution and can live in most conditions. Finally, we observed only two crayfish, animals that are somewhat sensitive to pollution.

CHEMICAL TESTS

Although physical tests cover areas seen with the naked eye, chemical tests can uncover pollutants that are not so recognizable. Certain chemicals can wipe out all life in a creek. Other chemicals can cause an overabundance of one life-form, which in turn could kill more sensitive animals.

A chart of results of chemical tests is included in the Test Comparison section on page 9. The chemical tests that D-Lynn performed were pH, dissolved oxygen (DO), turbidity, and phosphate.

■ **Figure 7–2** ■ continued

8

pH

The pH test is a measure of active hydrogen ions in a sample. The range of the pH test is 0–14. If the sample is in the range of 0–7.0, it is acidic; but if the sample is in the range of 7.0–14, it is basic. By law, the pH of a water sample must be within the range of 6.0–8.5.

For the tests we completed, the water sample was always 7.0, which is very good for a creek.

Dissolved Oxygen (DO)

Normally, oxygen dissolves readily into water from surface air. Once dissolved, it diffuses slowly in the water and is distributed throughout the creek. The amount of DO depends on different circumstances. Oxygen is always highest in choppy water, just after noon, and in cooler temperatures.

In many streams, the level of DO can become critically low during the summer months. When the temperature is warm, organisms are highly active and consume the oxygen supply. If the amount of DO drops below 3.0 ppm (parts per million), the area can become stressful for the organisms. An amount of oxygen that is 2.0 ppm or below will not support fish. DO that is 5.0 ppm to 6.0 ppm is usually required for growth and activity of organisms in the water.

According to the Water Quality Criteria for Georgia, average daily amounts of DO should be 5.0 ppm with a minimum of 4.0 ppm. Wildwood Creek scored well on this test. The average amount of DO in the water was 6.9 ppm, with the highest amount being 9.0 ppm on November 11, 2010.

Turbidity

Turbidity is the discoloration of water due to sediment, microscopic organisms, and other matter. One major factor of turbidity is the level of rainfall before a test.

Three of our tests were performed on clear days with little rainfall. On these dates, the turbidity of Wildwood Creek was always 1.0, the best that creek water can score on the test. The fourth test, which scored worse, occurred during a rainy period.

Phosphate

Phosphorus occurs naturally as phosphates—for example, orthophosphates and organically bound phosphates. Orthophosphates are phosphates that are formed in fertilizer, whereas organically bound phosphates can form in plant and animal matter and waste.

Phosphate levels higher than 0.03 ppm contribute to an increase in plant growth. If phosphate levels are above 0.1 ppm, plants may be stimulated to grow out of control. The phosphate level of Wildwood was always 0.5 ppm, considerably higher than is desirable.

■ **Figure 7–2** ■ continued

9

TEST COMPARISON

There was little change from each of the four test dates. The only tests that varied greatly from one test to another were air temperature, water temperature, water flow, and DO. On the basis of these results, it would appear that Wildwood Creek is a relatively stable environment.

Brings together test results for easy reference.

Table 1 Physical Tests

TEST DATES	5/26/10	6/25/10	9/24/10	11/19/10
Air Temperature in °C	21.5	23.0	24.0	13.5
Water Temperature in °C	20.0	22.0	23.0	13.0
Water Flow	Normal	High	Normal	Normal
Water Appearance	Clear	Clear	Clear	Clear
Habitat Description				
Number of Pools	2.0	3.0	2.0	5.0
Number of Ripples	1.0	2.0	2.0	2.0
Amount of Sediment Deposit	Average	Average	Good	Average
Stream Bank Stability	Average	Good	Good	Good
Stream Cover	Excellent	Good	Excellent	Good
Algae Appearance	Brown	Brown/hairy	Brown	Brown
Algae Location	Everywhere	Everywhere	Attached	Everywhere
Visible Litter	Heavy	Heavy	Heavy	Heavy
Bug Count	Low	Average	Low	Average

Table 2 Chemical Tests

Test	5/26/10	6/25/10	9/24/10	11/19/10
PH 7.0	7.0	7.0	7.0	
Dissolved Oxygen (DO)	6.8	6.0	5.6	9.0
Turbidity	1.0	3.0	1.0	1.0
Phosphate	0.50	0.50	0.50	0.50

■ **Figure 7–2** ■ continued

10

CONCLUSIONS AND RECOMMENDATIONS

This section includes the major conclusions and recommendations from our study of Wildwood Creek.

CONCLUSIONS

Generally, we were pleased with the health of the stream bank and its floodplain. The area studied has large amounts of vegetation along the stream, and the banks seem to be sturdy. The floodplain has been turned into a park, which handles floods in a natural way. Floodwater in this area comes in contact with vegetation and some dirt. Floodwater also drains quickly, which keeps sediment from building up in the creek.

However, we are concerned with the number and types of animals uncovered in our bug counts. Only two bug types were discovered, and these were types quite tolerant to pollutants. The time of year these tests were performed could affect the discovery of some animals. However, the low count still should be considered a possible warning sign about water quality. Phosphate levels were also high and probably are the cause of the large amount of algae.

We believe something in the water is keeping sensitive animals from developing. One factor that affects the number of animals discovered is the pollutant problems in the past (see Appendix A). The creek may still be in a redevelopment stage, thus explaining the small numbers of animals.

RECOMMENDATIONS

On the basis of these conclusions, we recommend the following actions for Wildwood Creek:

1. Conduct the current tests two more times, through Spring 2011. Spring is the time of year that most aquatic insects are hatched. If sensitive organisms are found then, the health of the creek could be considered to have improved.
2. Add testing for nitrogen. With the phosphate level being so high, nitrogen might also be present. If it is, then fertilizer could be in the water.
3. Add testing for human waste. Some contamination may still be occurring.
4. Add testing for metals, such as mercury, that can pollute the water.
5. Add testing for runoff water from drainage pipes that flow into the creek.
6. Schedule a volunteer cleanup of the creek.

With a full year of study and additional tests, the problems of Wildwood Creek can be better understood.

Draws conclusions that flow from data in body of report.

Uses paragraph format instead of lists because of lengthy explanations needed.

Gives numbered list of recommendations for easy reference.

■ **Figure 7–2** ■ continued

11

APPENDIX A

Background on Wildwood Creek

Wildwood Creek begins from tributaries on the northeast side of the city of Winslow. From this point, the creek flows southwest to the Chattahoochee River. Winslow Wastewater Treatment Plant has severely polluted the creek in the past with discharge of wastewater directly into the creek. Wildwood became so contaminated that signs warning of excessive pollution were posted along the creek to alert the public.

Today, all known wastewater discharge has been removed. The stream's condition has dramatically improved, but nonpoint contamination sources continue to lower the creek's water quality. Nonpoint contamination includes sewer breaks, chemical dumping, and storm sewers.

Another problem for Wildwood Creek is siltration. Rainfall combines with bank erosion and habitat destruction to wash excess dirt into the creek. This harsh action destroys most of the macroinvertebrates. At the present time, Wildwood Creek may be one of the more threatened creeks in Ware County.

■ **Figure 7–2** ■ continued

12

APPENDIX B

Water Quality Criteria for Georgia

All waterways in Georgia are classified in one of the following categories: fishing, recreation, drinking, and wild and scenic. Different protection levels apply to the different uses. For example, the protection level for dissolved oxygen is stricter in drinking water than fishing water. All water is supposed to be free from all types of waste and sewage that can settle and form sludge deposits.

In Ware County, all waterways are classified as "fishing," according to Chapter 391-3-6.03 of "Water Use Classifications and Water Quality Standards" in the Georgia Department of Natural Resources *Rules and Regulations for Water Quality Control.* The only exception is the Chattahoochee River, which is classified as "drinking water supply" and "recreational."

■ **Figure 7–2** ■ continued

13

APPENDIX C

Map 6
Location of City of Winslow
Parks and Recreation Facilities

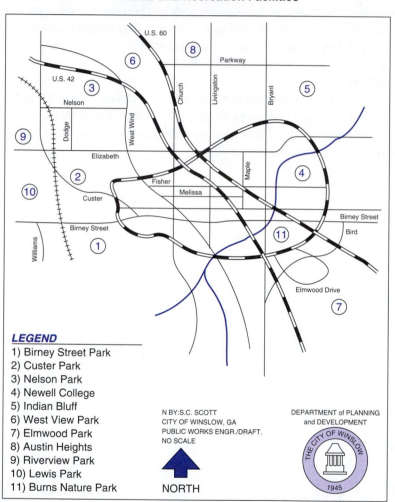

LEGEND
1) Birney Street Park
2) Custer Park
3) Nelson Park
4) Newell College
5) Indian Bluff
6) West View Park
7) Elmwood Park
8) Austin Heights
9) Riverview Park
10) Lewis Park
11) Burns Nature Park

N BY:S.C. SCOTT
CITY OF WINSLOW, GA
PUBLIC WORKS ENGR./DRAFT.
NO SCALE

NORTH

DEPARTMENT of PLANNING
and DEVELOPMENT

THE CITY OF WINSLOW
1945

■ **Figure 7–2** ■ continued

environmental problems. The paid professionals include engineers, environmental specialists, accountants, city planners, managers, lawyers, real estate experts, and public relations specialists. The part-time appointees include citizens who work in a variety of blue-collar and white-collar professions or who are homemakers.

>>> Four Common Reports

Report types vary from company to company and can be produced in either informal or formal report format. The four types described here are only a sampling of what you will be asked to write on the job. If you master these, you can probably handle other types that come your way.

Equipment Evaluations

All organizations use some kind of equipment that someone has to buy, maintain, or replace. Because companies put so much money into this part of their business, evaluating equipment is an important activity. Equipment evaluations provide objective data about how machinery, tools, computer software, or other equipment has functioned. An equipment evaluation may focus only on problems; or it may go on to suggest a change in equipment. Whatever its focus, an equipment evaluation must provide a well-documented review of the exact manner in which equipment performed.

mytechcommlab

Report 14: Research Report is an example of a formal equipment evaluation report.

Progress/Periodic Reports

Some reports are intended to cover activities that occurred during a specific period of time. They provide managers or clients with details about work on a specific project. A special type of progress report is the *project completion report*, which may be presented as a formal report when the project is completed. Progress and periodic reports contain mostly objective data. Yet both of them, especially progress reports, sometimes may be written in a persuasive manner. After all, you are trying to put forth the best case for the work you have completed. Figure 7–1 is an example of a progress report.

mytechcommlab

Report 12: Final Report is an example of a formal project completion report.

Problem Analyses

Every organization faces both routine and complex problems. Routine problems are often handled without much paperwork; they are discussed and then solved. However, other problems must often be described in reports, particularly if they involve many people, are difficult to solve, or have been brewing for a long time. Problem analysis reports present readers with a detailed description of problems in areas, such as personnel, equipment, products, and services.

mytechcommlab

Report 5: Research Report is an example of a problem analysis.

Most problem analyses contain both facts and opinions. As the writer, you must make special efforts to separate the two because most readers want the opportunity to draw their own conclusions about the problem.

Recommendation Reports

Recommendation reports use objective information to support suggestions that affect personnel, equipment, procedures, products, and services.

mytechcommlab

Report 6: Report of Findings is an example of a recommendation report.

>>> Chapter Summary

This chapter deals with the common types of reports you will write in your career. When writing reports, follow these guidelines:

1. Plan well before you write.
2. Separate fact from opinion.
3. Make text visually appealing.
4. Use illustrations for clarification and persuasion.
5. Edit carefully.

On the job, you write informal reports for readers inside your organization (as *memo reports*) and outside your organization (as *letter reports*). In both cases, follow these basic guidelines:

1. Use letter or memo format.
2. Use the ABC format for organizing information.
3. Start with an introductory summary.
4. Put detailed support in the body.
5. Focus attention in your conclusion.
6. Use attachments for details.

You also will write formal reports for large and complex projects, either inside or outside your organization, for readers with mixed technical backgrounds. Although long report formats vary according to company and profession, most have these nine basic parts: (1) cover/title page, (2) letter/memo of transmittal, (3) table of contents, (4) list of illustrations, (5) executive summary, (6) introduction, (7) discussion sections, (8) conclusions and recommendations, and (9) end material. Follow the specific guidelines in this chapter for these sections. The annotated example can serve as your reference.

Although formal and informal reports come in many varieties, this chapter describes only four common types: equipment evaluations, problem analyses, progress/periodic reports, and recommendations.

>>> **Learning Portfolio**

Collaboration at Work   Suggestions for High School Students

General Instructions

Each Collaboration at Work exercise applies strategies for working in teams to chapter topics. The exercise assumes you (1) have been divided into teams of about three to six students, (2) will use team time inside or outside of class to complete the case, and (3) will produce an oral or written response. For guidelines about writing in teams, refer to Chapter 2.

Background for Assignment

This chapter introduced you to five types of informal reports, one of which is focused on recommendations. When you are writing a recommendation report with a team of colleagues, agreeing on content can be a challenge. Recommendations involve opinions, and opinions often vary about what should be presented to the reader. As with other collaborative efforts, first you share information in a nonjudgmental way; then you choose what should be included in the report based on your team discussions.

Team Assignment

Assume an association of colleges and universities has asked your team to help write a short report to be sent to high school students. The report's purpose is to assist students in selecting a college or university. Your team will prepare an outline for the body of the report by (1) choosing several headings that classify groupings of recommendations and (2) providing specific recommendations within each grouping. For example, one grouping might be "Support for Job Placement," with one recommendation in this grouping being "Request data on the job placement rates of graduates of the institution." After producing your outline, share the results with other teams in the class.

Assignments

This chapter includes both short and long assignments. The short assignments in Part 1 are designed to be used for in-class exercises and short homework assignments. The assignments in Part 2 generally require more time to complete.

Part 1: Short Assignments

1. Problem Analysis—Critiquing a Report

Using the guidelines in this chapter, analyze the level of effectiveness of the problem analysis that begins on the following pages.

April 16, 2011
Mr Jay Henderson
Christ Church
10 Smith Dr
Jar Georgia 30060

**PROBLEM ANALYSIS:
NEW CHURCH BUILDING SITE**

Introductory Summary

Last week, your church hired our firm to study problems caused by the recent incorporation of the church's new building site into the city limits. Having reviewed the city's planning and zoning requirements, we have found some problems with your original site design—which initially was designed to meet the county's requirements only. My report focuses on problems with four areas on the site:

1. Landscaping screen
2. Church sign

(continued)

3. Detention pond
4. Emergency vehicle access

    Attached to this report is a site plan to illustrate these problems as you review the report. The plan was drawn from an aerial viewpoint.

Landscaping Screen

The city zoning code requires a landscaping screen along the west property line, as shown on the attached site illustration sheet. The former design does not call for a screen in this area. The screen will act as a natural barrier between the church parking lot and the private residence adjoining the church property. The code requires that the trees for this screen be a minimum height of 8 feet with a height maturity level of at least 20 feet. The trees should be an aesthetically pleasing barrier for all parties, including the resident on the adjoining property.

Church Sign

After the site was incorporated into the city, the Department of Transportation decided to widen Woodstock Road and increase the setback to 50 feet, as illustrated on our site plan. With this change, the original location of the sign falls into the road setback. Its new location must be out of the setback and moved closer to the new church building.

Detention Pond

The city's civil engineers reviewed the original site drawing and found that the detention pond is too small. If the detention pond is not increased, rainwater may build up and overflow into the building, causing a considerable amount of flood damage to property in the building and to the building itself. There is a sufficient amount of land in the rear of the site to enlarge and deepen the pond to handle all expected rainfall.

Emergency Vehicle Access

On the original site plan, the slope of the ground along the back side of the new building is so steep that an ambulance or city fire truck will not be able to gain access to the rear of the building in the event of a fire. This area is shown on our site illustration around the north and east sides of the building. The zoning office enforces a code that is required by the fire marshal's office. This code states that all buildings within the city limits must provide a flat and unobstructed access path around the buildings. If the access is not provided, the safety of the church building and its members would be in jeopardy.

Conclusion

The just-stated problems are significant, yet they can be solved with minimal additional cost to the church. Once the problems are remedied and documented, the revised site plan must be approved by the zoning board before a building permit can be issued to the contractor.

    I look forward to meeting with you and the church building committee next week to discuss any features of this study and its ramifications.

Sincerely,

Thomas K. Jones

Thomas K. Jones
Senior Landscape Engineer

Enclosure

2. Writing a Recommendation Report

Divide into teams of three or four students, as your instructor directs. Consider your team to be a technical team from an architectural engineering firm. Assume that the facilities director of your college or university has hired your team to recommend changes that would improve your classroom. Write a team report that includes the recommendations agreed to by your team. For example, you may want to consider structural changes of any kind, additions of equipment, changes in the type and arrangement of seating, and so forth.

3. Evaluation—A Formal Report

Locate a formal report written by a private firm or government agency, or use a long report provided by your instructor. Determine the degree to which the example follows the guidelines in this chapter. Depending on the instructions given by your teacher, choose between the following options:

- Present your findings orally or in writing.
- Select part of the report or all of the report.

5. Progress or Periodic Report

Assume that you receive financial support for college from an agency in a nearby large city that requires regular reports on your academic progress. Choose one of these two options for this assignment:

Progress report: Select a major project you are now completing in any college course. Following the guidelines in the "Progress/Periodic Reports" section of this chapter, write a progress report on this project. Direct the letter report to Wade Simkins, Financial Aid Director of Today's

Students for Tomorrow's Success. Sample topics might include a major paper, laboratory experiment, field project, or design studio.

Periodic report: Following the guidelines in the "Progress/ Periodic Reports" section of this chapter, write a periodic report on your recent course work (completed or ongoing classes or both). Direct the letter report to Wade Simkins, Financial Aid Director of Today's Students for Tomorrow's Success. Organize the report by class and then give specific updates on each one.

Part 2: Longer Assignments—Individual or Team Work

This section contains assignments for writing longer informal reports and entire formal reports. Remember to complete the Planning Form for each assignment. These assignments can be written by individual writers or by writing teams. If your instructor has made this a team assignment, review the guidelines on team writing in Chapter 2.

6. Problem Analysis

Assume that you are a landscape engineer whose company examines problems associated with the design of walkways, the location of trees and garden beds, the grading of land around buildings, and any other topographical features. You have been hired by a specific college, community, or company with which you are familiar. Your objective is to evaluate one or more landscaping problems at the site.

Write an informal report that describes the problem(s) in detail. (Follow the guidelines in the "Problem Analyses" section of this chapter.) Be specific about how the problem affects people—the employees, inhabitants, students, and so forth. Following are some sample problems that could be evaluated:

- Poorly landscaped entrance to a major subdivision
- Muddy, unpaved walkway between dormitories and academic buildings on a college campus
- Unpaved parking lot far from main campus buildings
- Soil runoff into the streets from several steep, muddy subdivision lots that have not yet been sold
- City tennis courts with poor drainage
- Lack of adequate flowers or bushes around a new office building
- Need for a landscaped common area within a subdivision or campus
- Need to save some large trees that may be doomed because of proposed construction

7. Recommendation Report

This project requires some research. Assume that your college plans either to embark on a major recycling effort or to expand a recycling program that has already started. Put yourself in the role of an environmental scientist or technician who has been asked to recommend these recycling changes.

First, do some research about recycling programs that have worked in other organizations. A good place to start is a periodical database such as *EBSCOhost* or *J-Stor*, which will lead you to some magazine articles of interest. Choose to discuss one or more recoverable resources, such as paper, aluminum, cardboard, plastic, or glass bottles. Be specific about how your recommendations can be implemented by the organization or audience about which you are writing. (Consult the guidelines in the "Recommendation Reports" section of this chapter.)

mytechcommlab

For more practice, do the Revising a Formal Report on Technology Activity.

8. Research-Based Formal Report

Complete the following procedure for writing a research-based report:

- Use library and Internet resources to research a general topic in a field that interests you. Do some preliminary reading to screen possible specific topics.
- Choose three to five specific topics that require further research and for which you can locate information.
- Work with your instructor to select the one topic that best fits this assignment, given your interests and the criteria set forth here.
- Develop a simulated context for the report topic, whereby you select a purpose for the report, a specific audience to whom it could be addressed (as if it were a real report), and a specific role for you as a writer.

For example, assume you have selected "Earth-Sheltered Homes" as your topic. You might be writing a report to the manager of a local design firm on the features and construction techniques of such structures. As a newly hired engineer or designer, you are presenting information so that your manager can decide whether the firm might want to begin building and marketing such homes. This report might present only data, or it could present data and recommendations.

Write the report according to the format guidelines in this chapter and in consideration of the specific context you have chosen.

9. Ethics Assignment

Illustrations on cover pages of formal reports are one strategy for attracting the readers' attention to the document. Note the use of an illustration on page 159. Noland Engineering submitted a formal report to a coastal city in California, concluding that an industrial park can be built near the city's bird sanctuary without harming the habitat—if stringent guidelines are followed. The report writer decided to place the picture of a bird on the title page, punctuating the report's point about the industrial park. Do you think the report cover uses its graphics in an ethically sound way to engage the reader with the report? Why or why not? How do you determine whether a cover page illustration is an appropriate persuasive tool on the one hand, or an inappropriate attempt to manipulate the reader on the other hand? Give hypothetical examples, or find examples from reports available on the Internet.

10. International Communication Assignment

This assignment requires that you gain information about writing long technical reports designed for readers outside the United States (or outside the country where you are taking this course, if it is not the United States). The suggestions you develop can relate to either (1) reports written in English that will be read in English or (2) reports written in English that will be translated into another language.

Specifically, write a report that provides a wide range of recommendations for writing long technical documents to a specific international audience. Cover as many writing-related issues as possible—organization, format, page design, and style.

To gather information for this assignment, find someone who works for an international firm, deals with international clients, or has in some other way acquired information about the needs of international readers of technical documents. Possible sources include (1) your institution's alumni office, which may be able to provide names of graduates or employers of graduates; (2) friends or colleagues; (3) individuals contacted through Web sites of international organizations; and (4) local chambers of commerce and other organizations that promote international trade.

Oceanside's New Industrial Park

Prepared for: City Council
                 Oceanside, California

Prepared by: Noland Engineering, L.L.P.
                 San Francisco, California

Date: March 3, 2010

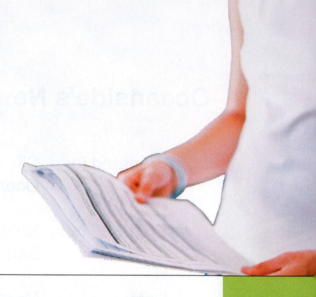

Chapter | 8 | Proposals

Everyone of us relies on persuasion. Usually we apply our persuasive strategies in less formal situations—for example, in asking for an extension on a paper deadline, gaining approval to use a sales coupon after its void date, convincing a friend to choose one vacation destination over another, or selling retail products on commission. When persuasion is applied to business situations, it often results in a written document called a *proposal*, and you may write many of them in your career. Proposals are defined as follows:

> **Proposal:** A document written to convince your readers to adopt an idea, a product, or a service. They can be directed to colleagues inside your own organization (*in-house proposals*), to clients outside your organization (*sales proposals*), or to organizations that fund research and other activities (*grant proposals*).
>
> In all three cases, proposals can be presented in either a short, simple format (*informal proposal*) or a longer, more complicated format (*formal proposal*). Also, proposals can be either requested by the reader (*solicited*) or submitted without a request (*unsolicited*).

Like informal reports, informal proposals are short documents that cover projects with a limited scope. Following are some guidelines to help you decide when to use informal and formal proposals:

Use Informal Proposals When

- The text of the proposal (excluding attachments) is no more than five pages
- The size of the proposed project is such that a long formal proposal appears to be inappropriate
- The client has expressed a preference for a leaner and less-formal document

Use Formal Proposals When

- The text of the proposal (excluding attachments) is more than five pages
- The size and importance of the project is such that a formal proposal is appropriate
- The client has expressed a preference for a more-formal document

These two formats can be used for proposals that are either *in-house* (to readers within your own organization) or *external* (to readers outside your organization).

The flowchart in Figure 8–1 shows one possible communication cycle that would involve both a proposal and an *RFP*, which stands for *request for proposal*.

> **Request for proposal (RFP):** A document sometimes sent out by organizations that want to receive proposals for a product or service. The RFP gives guidelines on (1) what the proposal should cover, (2) when it should be submitted, and (3) to whom it should be sent. As writer, you should follow the RFP religiously in planning and drafting your proposal.

RFPs generally are used when an organization:

- Wants to receive multiple approaches to address a problem.
- Wants the best price for a well-defined scope of work.
- Is required to solicit bids for projects.
- Provides grants or fellowships for research or community projects.

If you are responding to an RFP, it is key to your understanding the needs of the readers to whom you will send your proposal. You must read RFPs very carefully to understand the main concerns and interests of your readers, as well as the requirements for proposals. You should precisely follow the content and formatting guidelines outlined in the RFP, as proposals that do not follow the guidelines are usually rejected immediately.

■ Figure 8–1 ■

Flowchart showing the main documents involved in an external, solicited proposal process.

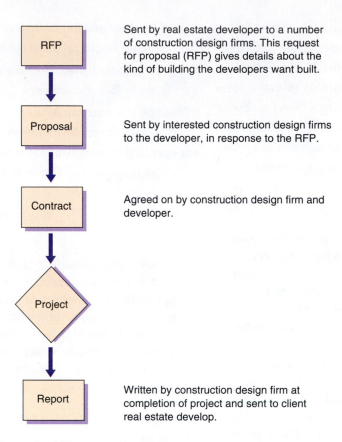

RFP — Sent by real estate developer to a number of construction design firms. This request for proposal (RFP) gives details about the kind of building the developers want built.

Proposal — Sent by interested construction design firms to the developer, in response to the RFP.

Contract — Agreed on by construction design firm and developer.

Project

Report — Written by construction design firm at completion of project and sent to client real estate develop.

>>> Guidelines for Informal Proposals

This section provides writing guidelines and an annotated model for informal proposals, which have two formats: (1) memos (for in-house proposals) and (2) letters (for external proposals). With some variations, these guidelines are similar to those suggested in Chapter 7 for informal reports. The formats are much the same, although the content and tone are different. Reports *explain,* whereas proposals *persuade.*

>> Informal Proposal Guideline 1: Plan Well Before You Write

Complete the Planning Form at the end of the book for all proposal assignments. Carefully consider your purpose, audience, and organization. Two factors make this task especially difficult in sales-proposal writing:

1. You may know nothing more about the client than what is written on the RFP.

2. Proposals often are on a tight schedule, which limits your planning time.

Nevertheless, try to find out exactly who will be making the decision about your proposal. Many clients will tell you if you give them a call. In fact, they may be pleased that you care enough about the project to target the audience. Once you identify the decision makers, spend time brainstorming about their needs before you begin writing.

>> Informal Proposal Guideline 2: Use Letter or Memo Format

Letter proposals usually follow the format of routine business letters (see Chapter 4). This casual style gives readers the immediate impression that your document will be *approachable*—that is, easy to get through and limited in scope. Memo proposals, such as the example shown in Figure 8–2, follow the format of an internal memorandum (see Chapter 4). Following are a few highlights:

- Line spacing is usually single, but it may be one-and-a-half or double, depending on the reader's or company's preference.
- The recipient's name, date, and page number appear on sheets after the first.
- Most readers prefer an uneven or ragged-edge right margin, as opposed to an even or full-justified margin.

Your subject line in a memo proposal gives readers the first impression of the proposal's purpose. Choose concise yet accurate wording. See Figure 8–2 for wording that gives the appropriate information and tries to engage the reader's interest.

>> Informal Proposal Guideline 3: Make Text Visually Appealing

The page design of informal proposals must draw readers into the document. Remember—you are trying to sell a product, a service, or an idea. Also, remember that your proposal may be competing with others. Put yourself in the place of the reader who is wondering which one to pick up first. How the text looks on the page can make a big difference. Following are a few techniques to follow to help make your proposal visually appealing:

- Use lists (with bullets or numbered points) to highlight main ideas.
- Follow your readers' preferences as to font size, type, line spacing, and so forth. Proposals written in the preferred format of the reader gain an edge.
- Use headings and subheadings to break up blocks of text.

These and other techniques help guide readers through the informal proposal. Given that there is no table of contents, you must take advantage of such strategies.

DATE: October 3, 2010
TO: Gary Lane
FROM: Jeff Bilstrom **JB**
SUBJECT: Creation of Logo for Montrose Service Center

Gives concise view of problem—*and* his proposed solution.

Part of my job as director of public relations is to get the Montrose name firmly entrenched in the minds of metro Atlanta residents. Having recently reviewed the contacts we have with the public, I believe we are sending a confusing message about the many services we offer retired citizens in this area.

To remedy the problem, I propose we adopt a logo to serve as an umbrella for all services and agencies supported by the Montrose Service Center. This proposal gives details about the problem and the proposed solution, including costs.

The Problem

Includes effective lead-in.

The lack of a logo presents a number of problems related to marketing the center's services and informing the public. Here are a few:

Uses bulleted list to highlight main difficulties posed by current situation.

- The letterhead mentions the organization's name in small type, with none of the impact that an accompanying logo would have.
- The current brochure needs the flair that could be provided by a logo on the cover page, rather than just the page of text and headings that we now have.
- Our 14 vehicles are difficult to identify because there is only the lettered organization name on the sides without any readily identifiable graphic.
- The sign in front of our campus, a main piece of free advertising, could better spread the word about Montrose if it contained a catchy logo.
- Other signs around campus could display the logo, as a way of reinforcing our identity and labeling buildings.

Ends section with good *transition* to next section.

It is clear that without a logo, the Montrose Service Center misses an excellent opportunity to educate the public about its services.

The Solution

Starts with *main point*—need for logo.

I believe a professionally designed logo could give the Montrose Service Center a more distinct identity. Helping to tie together all branches of our operation, it would give the public an easy-to-recognize symbol. As a result, there would be a stronger awareness of the center on the part of potential users and financial contributors.

■ **Figure 8–2** ■ Memo proposal

Gary Lane
October 3, 2010
Page 2

The new logo could be used immediately to do the following:

- Design and print letterhead, envelopes, business cards, and a new brochure. ◄──── *Focuses on benefits of proposed change.*
- Develop a decal for all company vehicles that would identify them as belonging to Montrose.
- Develop new signs for the entire campus, to include a new sign for the entrance to the campus, one sign at the entrance to the Blane Workshop, and one sign at the entrance to the Administration Building.

Cost

Developing a new logo can be quite expensive. However, I have been able to get the name of a well-respected graphic artist in Atlanta who is willing to donate his ◄──── *Emphasizes benefit of possible price break.* services in the creation of a new logo. All that we must do is give him some general guidelines to follow and then choose from among eight to ten rough sketches. Once a decision is made, the artist will provide a camera-ready copy of the new logo.

• Design charge	$0.00
• Charge for new letterhead, envelopes, business cards, and brochures (min. order)	545.65
• Decal for vehicles 14 @ $50.00 + 4%	728.00
• Signs for campus	415.28
Total Cost	$1,688.93

Uses listing to clarify costs.

Conclusion ◄────

Closes with major benefit to reader and urge to action.

As the retirement population of Atlanta increases in the next few years, there will be a much greater need for the services of the Montrose Service Center. Because of that need, it is in our best interests to keep this growing market informed about the organization.

I'll stop by later this week to discuss any questions you might have about this ◄──── *Keeps control of next step.* proposal.

■ **Figure 8–2** ■ continued

ABC Format:

Informal Proposal

- **ABSTRACT:** Gives "big picture" for those who make decisions. Usually includes a statement of the problem or other information that will entice the audience to read further.
- **BODY:** Gives details about exactly what you are proposing to do.
- **CONCLUSION:** Drives home the main benefit and makes clear the next step.

>> Informal Proposal Guideline 4: Use the ABC Format for Organization

The ABC Format used throughout this book also applies to informal proposals.

Note: Beginning and end sections should be easy to read and stress just a few points. They provide a short buffer on both ends of the longer and more technical body section in the middle. The next four guidelines give more specific advice for writing the main parts of an informal proposal.

>> Informal Proposal Guideline 5: Create the Abstract as an Introductory Summary

In an abstract, you capture the client's attention with a capsule summary of the entire proposal. This is especially important in unsolicited proposals. This one- or two-paragraph starting section permits space only for what the reader really must know at the outset, such as the following:

- Purpose of proposal
- Reader's main need
- Main features you offer, as well as related benefits
- Overview of proposal sections to follow

Keep this overview very brief. Answer the one question readers are thinking: "Why should I accept this proposal?"

>> Informal Proposal Guideline 6: Put Important Details in the Body

The discussion of your proposal should address these basic questions:

1. What problems are you trying to solve, and why?
2. What are the technical details of your approach?
3. Who will do the work, and with what?
4. When will it be done?
5. How much will it cost?

Discussion formats vary from proposal to proposal, but here are some sections commonly used to respond to these questions:

1. **Description of problem or project and its significance.** Give a precise technical description, along with any assumptions you have made on the basis of previous

contact with the reader. Explain the importance or significance of the problem to the reader of the proposal.

2. **Proposed solution or approach.** Describe specific tasks you propose in a manner that is clear and well organized. If you are presenting several options, discuss each one separately—making it easy for the reader to compare and contrast information.

3. **Personnel.** If the proposal involves people performing tasks, it may be appropriate to explain qualifications of participants.

4. **Schedule.** Even the simplest proposals require some information about the schedule for delivering goods, performing tasks, and so forth. Be both clear and realistic in this portion of the proposal. Use graphics when appropriate.

5. **Costs.** Place complete cost information in the body of the proposal unless you have a table that would be more appropriately placed in an attachment. Above all, do not bury dollar figures in paragraph format. Instead, highlight figures with indented or bulleted lists. Because your reader will be looking for cost data make that information easy to find. Finally, be certain to include all costs—materials, equipment, personnel, salaries, and so forth.

>> Informal Proposal Guideline 7: Use a Problem/Solution Organization

As any good salesperson knows, customers must feel that they need your product, service, or idea before they can be convinced to purchase or support it. Lay the groundwork for acceptance by first showing the readers that a strong need exists.

Establishing need is most crucial in unsolicited proposals, of course, when readers may not be psychologically prepared to accept a change that costs money. Even in proposals that have been solicited, however, you should give some attention to restating the basic needs of the readers. You should show your readers your understanding of the problem, and you should emphasize the connections between the problem and your proposed solution.

> **mytechcommlab**
> The Problem/Solution pattern is clear in Research Report 2.

>> Informal Proposal Guideline 8: Focus Attention in Your Conclusion

Called *conclusion* or *closing,* this section gives you the opportunity to control the readers' last impression. It also helps avoid the awkwardness of ending proposals with the costs. In this closing section, you can

- Emphasize a main benefit or feature of your proposal
- Restate your interest in doing the work
- Indicate what should happen next

Regarding the last point, sometimes you may ask readers to call if they have questions. In other situations, however, it is appropriate to say that you will follow up the proposal with a phone call. This approach leaves you in control of the next step.

>> **Informal Proposal Guideline 9:** Use Attachments for Less Important Details

The text of informal proposals is usually less than five pages, so you may have to place supporting data or illustrations in attachments that follow the conclusion. Cost and schedule information, in particular, is best placed at the end in well-labeled sections.

Make sure the proposal text includes clear references to all visuals. If you have more than one attachment, give each one a letter and a title (for example, "Attachment A: Project Costs"). If you have only one attachment, include the title but no letter (for example, "Attachment: Resumes").

>> **Informal Proposal Guideline 10:** Edit Carefully

Build in enough time for a series of editing passes, preferably by different readers. There are two reasons why proposals of all kinds deserve this special attention.

1. They can be considered contracts in a court of law. If you make editing mistakes that alter meaning (such as an incorrect price figure), you could be bound to the error.
2. Proposals often present readers with their first impression of you. If the document is sloppy, they can make assumptions about your professional abilities as well.

>>> Guidelines for Formal Proposals

Sometimes the complexity of the proposal may be such that a formal response is best. Ask yourself questions such as the following in deciding whether to write an informal or a formal proposal:

- Is there too much detail for a letter or memo?
- Is a table of contents needed so that sections can be found quickly?
- Will the professional look of a formal document lend support to the cause?
- Are there so many attachments that a series of lengthy appendices would be useful?
- Are there many different readers with varying needs, such that there should be different sections for different people?

If you answer "yes" to one or more of these questions, you probably should write a formal proposal. Although this long format is most common in external sales proposals, some in-house proposals may require the same approach—especially in large organizations in which you may be writing to unknown persons in distant departments.

Formal proposals can be long and complex, so this part of the chapter treats each proposal section separately—from title page through conclusion. Two points will become evident as you use these guidelines. First, formal reports and formal proposals are very much alike. A quick look at Chapter 7 shows you the similarities in format. Second, a formal proposal follows the basic ABC format described in Chapter 1.

Specifically, the parts of the formal proposal fit the pattern as shown in the following box. Notice that formal proposals, like informal proposals, are organized with a problem/solution pattern.

As you read through and apply these guidelines, refer to Figure 8–3 on pp. 174–182 for an annotated example of the formal proposal.

Cover/Title Page

Like formal reports, formal proposals are usually bound documents with a cover, which includes one or more of the items listed on the right for inclusion on the title page. The cover should be designed to attract the reader's interest—with good page layout and perhaps even a graphic.

Inside the cover is the title page, which contains these four pieces of information:

- **Project title,** sometimes preceded by *Proposal for* or similar wording
- **Your reader's name** (sometimes preceded by *Prepared for...*)
- **Your name or the name of your organization** spelled out in full (sometimes preceded by *Prepared by...*)
- **Date of submission**

> ## ABC Format: **Formal Proposal**
>
> - **ABSTRACT:**
>   - Cover/Title Page
>   - Letter of Transmittal
>   - Table of Contents
>   - List of Illustrations
>   - Executive Summary
>   - Introduction
> - **BODY:**
>   - Technical information
>   - Management information
>   - Cost information
>   - (Appendices—appear after text, but support Body section)
> - **CONCLUSION:**
>   - Conclusion

The title page gives clients their first impression of you. For that reason, consider using some tasteful graphics to make the proposal stand out from those of your competitors.

Letter/Memo of Transmittal

Internal proposals have *memos of transmittal;* external sales proposals have *letters of transmittal.* The guidelines for format and organization presented here help you write attention-getting prose. In particular, note that the letter or memo should be in single-spaced, ragged-right-edge format, even if the rest of the proposal is double-spaced copy with right-justified margins.

For details of letter and memo format, see Chapter 4. The guidelines for the letter/memo of transmittal for formal reports also apply to formal proposals (see "Transmittal Guideline 4" in Chapter 7). For now, here are some highlights of format and content that apply especially to letters and memos of transmittal:

1. Use short beginning and end paragraphs.
2. Use a conversational style, with little or no technical jargon.

3. Use the first paragraph for introductory information, mentioning what your proposal responds to (e.g., a formal RFP, a conversation with the client, your perception of a need).

4. Use the middle of the letter to emphasize one main benefit of your proposal. Stress what you can do to solve a problem, using the words *you* and *your* as much as possible (rather than *I* and *we*).

5. Use the last paragraph to retain control by orchestrating the next step in the proposal process. When appropriate, indicate that you will call the client soon to follow up on the proposal.

6. Follow one of the letter formats described in Chapter 4. Following are some exceptions, additions, or restrictions:

   ■ Use single-spaced, ragged-right-edged copy, which makes your letter stand out from the proposal proper.
   ■ Keep the letter on one page—a two-page letter loses that crisp and concise impact you want to make.
   ■ Place the company proposal number (if there is one) at the top, above the date. Exact placement of both number and date depends on your organization's letter style.
   ■ Include the client's company name or personal name on the first line of the inside address, followed by the mailing address used on the envelope. Include the full name (and title, if appropriate) of your contact person at the client firm. If you use a personal name, follow the last line of the inside address with a conventional greeting ("Dear Mr. Adams:").
   ■ (Optional) Include the project title beneath the attention line, using the exact wording that appears on the title page.
   ■ Close with "Sincerely" and your name at the bottom of the page. Also include your company affiliation.

Table of Contents

Create a very readable table of contents by spacing items well on the page. List all proposal sections, subsections, and their page references. At the end, list any appendices that may accompany the proposal.

Given the tight schedule on which most proposals are produced, errors can be introduced at the last minute because of additions or revisions. Therefore, take time to proofread the table of contents carefully. In particular, make sure to follow these guidelines:

■ Wording of headings should match within the proposal text.
■ Page references should be correct.
■ All headings of the same order should be parallel in grammatical form.

List of Illustrations

When there are many illustrations, the list of illustrations appears on a separate page after the table of contents. When there are few entries, however, the illustrations may be listed at the end of the table of contents page. In either case, the list should include the number, title, and page number of every illustration appearing in the body of the text. (If there is only one illustration, a number need not be included.) You may divide the list into tables and figures.

Executive Summary

Executive summaries are the most frequently read parts of proposals. Often read by decision makers in an organization, the summary should present a concise overview (usually one page) of the proposal's most important points. It should also accomplish the following objectives:

- Avoid technical language.
- Be as self-contained as possible.
- Make brief mention of the problem, proposed solution, and cost.
- Emphasize the main benefits of your proposal.

Start the summary with one or two sentences that command readers' attention and engage their interest, and then focus on just a few main selling points (three to five is best). You might even want to highlight these benefits with indented lead-ins, such as *Benefit 1* and *Benefit 2*. When possible, use the statement of benefits to emphasize what is unique about your company or your approach so that your proposal attracts special attention. Finally, remember to write the summary after you have completed the rest of the proposal. Only at this point do you have the perspective to sit back and develop a reader-oriented overview.

Introduction

The introduction provides background information for both nontechnical and technical readers. Although the content varies from proposal to proposal, some general guidelines apply. You should include information on the (1) purpose, (2) description of the problem to which you are responding, (3) scope of the proposed study, and (4) format of the proposal.

- Use subheadings if the introduction goes over a page. In this case, begin the section with a lead-in sentence or two that mention the sections to follow.
- Start with a purpose statement that concisely states the reason you are writing the proposal.

■ Include a description of the problem or need to which your proposal is responding. Use language directly from the request for proposal or other document the reader may have given you so that there is no misunderstanding. For longer problem or need descriptions, adopt the alternative approach of including a separate needs section or problem description after the introduction.

■ Include a scope section in which you briefly describe the range of proposed activities covered in the proposal, along with any research or preproposal tasks that have already been completed.

■ Include a proposal format section if you believe the reader would benefit from a listing of the major proposal sections that follow.

Some formal sales proposals include information about the history, background, and expertise of an organization. This is especially appropriate in proposals that are responding to an RFP, where a number of organizations are competing to win the contract. This material is often *boilerplate,* or text that can be reused in all similar documents.

Discussion Sections

Aim the discussion or body toward readers who need supporting information. Traditionally, the discussion of a formal sales proposal contains three basic types of information: (1) technical, (2) management, and (3) cost. Following are some general guidelines for presenting each type. Remember that the exact wording of headings and subheadings varies depending on proposal content.

1. **Technical Sections**
   ■ Respond thoroughly to the client's concerns, as expressed in writing or meetings.
   ■ Follow whatever organization plan that can be inferred from the request for proposal.
   ■ Use frequent subheadings with specific wording.
   ■ Back up all claims with facts.

2. **Management Sections**
   ■ Describe who will do the work.
   ■ Explain when the work will be done.
   ■ Display schedule information graphically.
   ■ Highlight personnel qualifications (but put resumes in appendices).

3. **Cost Section**
   ■ Make costs extremely easy to find.
   ■ Use formal or informal tables when possible.
   ■ Emphasize value received for costs.

■ Be clear about add-on costs or options.

■ Always total your costs.

Conclusion

Formal proposals should always end with a section labeled "Conclusion" or "Closing." This final section of the text gives you the chance to restate a main benefit, summarize the work to be done, and assure clients that you plan to work with them closely to satisfy their needs. Just as important, this brief section helps you end on a positive note. You come back full circle to what you stressed at the beginning of the document—benefits to the client and the importance of a strong personal relationship. (Without the conclusion, the client's last impression would be made by the cost section in the discussion.)

Appendices

Because formal proposals are so long, readers sometimes have trouble locating information they need. Headings help, but they are not the whole answer. Another way you can help readers is by transferring technical details from the proposal text into appendices. The proposal still contains detail—for technical readers who want it—but details do not intrude into the text.

Proposals often include boilerplate, such as the resumes of all major personnel who will be working on a project or project sheets that show expertise and experience with projects like those on which the company is bidding. Creating boilerplate for this information can save you or your employer's considerable time by eliminating the need to recreate this material for each new proposal.

Any supporting information can be placed in appendices, but following are some common items included there as well:

■ Resumes

■ Organization charts

■ Company histories

■ Detailed schedule charts

■ Contracts

■ Cost tables

■ Detailed options for technical work

■ Summaries of related projects already completed

■ Questionnaire samples

This boilerplate is often printed from separate files, and thus is not paged in sequence with your text. Instead, it is best to use individual paging within each appendix. For example, pages in an Appendix B are numbered as B-1, B-2, B-3, and so forth.

**PROPOSAL FOR SUPPLYING
TEAK CAM CLEAT SPACERS**

Prepared by
Totally Teak, Inc.

Prepared for
John L. Riggini
Bosun's Locker Marine Supply

August 22, 2011

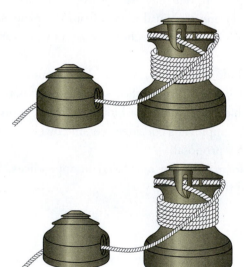

■ **Figure 8–3** ■ Formal proposal

Totally Teak Inc
6543 Amster Avenue NW
Atlanta Georgia 30308
404.555.9425

August 22, 2011

John L Riggini President
Bosun's Locker Marine Supply
38 Oakdale Parkway
Norcross OH 43293

Dear Mr. Riggini:

I enjoyed talking with you last week about inventory needs at the 10 Bosun's stores. In response to your interest in our products, I'm submitting this proposal to supply your store with our Teak Cam Cleat Spacers.

This proposal outlines the benefits of adding Teak Cam Cleat Spacers to your line of sailing accessories. The potential for high sales volume stems from the fact that the product satisfies two main criteria for any boat owner:

1. It enhances the appearance of the boat.
2. It makes the boat easier to handle.

Your store managers will share my enthusiasm for this product when they see the response of their customers.

I'll give you a call next week to answer any questions you have about this proposal.

Sincerely,

William G. Rugg

William G. Rugg
President
Totally Teak, Inc.

WR/rr

In this example, the letter of transmittal appears immediately following the title page. It can also appear before the title page (see Model 8–4).

Establishes *link* with previous client contact.

Stresses two main benefits.

Says he will *call* (rather than asking client to call).

■ **Figure 8–3** ■ continued

CONTENTS

Organizes entire proposal around *benefits*. → **FEATURES AND BENEFITS**

ILLUSTRATIONS

■ **Figure 8–3** ■ continued

EXECUTIVE SUMMARY

This proposal outlines features of a custom-made accessory designed for today's sailors—whether they be racers, cruisers, or single-handed skippers. The product, Teak Cam Cleat Spacers, has been developed for use primarily on the Catalina 22, a boat owned by many customers of the 10 Bosun's stores. However, it can also be used on other sailboats in the same class.

The predictable success of Teak Cam Cleat Spacers is based on two important questions asked by today's sailboat owners:

- Will the accessory enhance the boat's appearance?
- Will it make the boat easier to handle and, therefore, more enjoyable to sail?

This proposal answers both questions with a resounding affirmative by describing the benefits of teak spacers to thousands of people in your territory who own boats for which the product is designed. This potential market, along with the product's high profit margin, will make Teak Cam Cleat Spacers a good addition to your line of sailing accessories.

Briefly mentions main need to which proposal responds.

Reinforces main points mentioned in letter (selective repetition of crucial information is acceptable).

1

■ **Figure 8–3** ■ continued

Makes clear the proposal's purpose.

Gives lead-in about section to follow.

Establishes *need* for product.

Shows his understanding of *need* (personal experience of designing owner survey).

Gives list of sections to follow, to reinforce organization of proposal.

INTRODUCTION

The purpose of this proposal is to show that Teak Cam Cleat Spacers will be a practical addition to the product line at the Bosun's Locker Marine Supply stores. This introduction highlights the need for the product, as well as the scope and format of the proposal.

Background

Sailing has gained much popularity in recent years. The high number of inland impoundment lakes, as well as the vitality of boating on the Great Lakes, has spread the popularity of the sport. With this increased interest, more and more sailors have become customers for a variety of boating accessories.

What kinds of accessories will these sailors be looking for? Accessories that (1) enhance the appearance of their sailboats and (2) make their sailboats easier to handle and, consequently, more enjoyable to sail. With these customer criteria in mind, it is easy to understand the running joke among boat owners (and a profitable joke among marine supply dealers): "A boat is just a hole in the water that you pour your money into."

The development of this particular product originated from our designers' first-hand sailing experiences on the Catalina 22 and knowledge obtained during manufacture (and testing) of the first prototype. In addition, we conducted a survey of owners of boats in this general class. The results showed that winch and cam cleat designs are major concerns.

Proposal Scope and Format

The proposal focuses on the main advantages that Teak Cam Cleat Spacers will provide your customers. These six sections follow:

1. Practicality
2. Suitability for a Variety of Sailors
3. High-Quality Construction and Appearance
4. Dealer Benefits
5. Sizable Potential Market
6. Affordable Price

2

■ **Figure 8–3** ■ continued

FEATURES AND BENEFITS

Teak Cam Cleat Spacers offer Bosun's Locker Marine Supply the best of both worlds. On the one hand, the product solves a nagging problem for sailors. On the other hand, it offers your store managers a good opportunity for profitability. Described here are six main benefits for you to consider.

Uses main heading that *engages reader's interest.*

Practicality

This product is both functional and practical. When installed in the typical arrangement shown in the figure below, the Teak Cam Cleat Spacer raises the height of the cam cleat, thereby reducing the angle between the deck and the sheet as it feeds downward from the winch. As a result of this increased height, a crewmember is able to cleat a sheet with one hand instead of two.

Phrases each side heading in "benefit-centered" language.

SIDE VIEWS OF CLEATING ARRANGEMENT

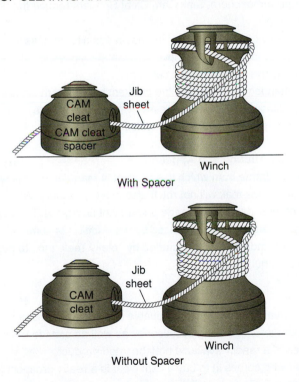

3

Explains *exactly how* product will work.

Such an arrangement allows a skipper to maintain steerage of the boat, keeping one hand on the helm while cleating the sheet with the other. Securing a sheet in this manner can be done more quickly and securely. Also, this installation reduces the likelihood of a sheet "popping out" of the cam cleat during a sudden gust of wind.

Moves logically from *racing* to *cruising* to *single-handed sailing*—all to show usefulness of product.

Suitability for a Variety of Sailors

For the racer, cruiser, and single-handed sailor, sailing enjoyment is increased as sheets and lines become easier to handle and more secure. In a tight racing situation, these benefits can be a deciding factor. The sudden loss of sail tension at the wrong moment as a result of a sheet popping out of the cam cleat could make the difference in a close race.

A cruising sailor is primarily concerned with relaxation and pleasure. A skipper in this situation wants to reduce his or her workload as much as possible. In the instance of a sheet popping loose, the sudden chaos of a sail flapping wildly interrupts an otherwise tranquil atmosphere. Teak Cam Cleat Spacers reduce the chance of this happening.

A cruising sailor often has guests abroad. In this situation, as well as in a race, the skipper wants to maintain a high level of seamanship, especially where the control of the boat and the trim of its sails are concerned.

The single-handed sailor derives the greatest benefit from installing Teak Cam Cleat Spacers. Without crew nearby to assist with handling lines or sheets, anything that makes work easier for the skipper is welcome.

Stresses *quality*.

High-Quality Construction and Appearance

The teakwood frame from which this product is manufactured is well suited for use around water, since teak will not rot. It also looks nice when oiled or varnished.

The deck of most sailboats is made primarily of fiberglass. The appearance of such a boat can be significantly enhanced by the addition of some teak brightwork.

Each spacer is individually handcrafted by Totally Teak, Inc., to guarantee a consistent level of high quality.

Appeals to self-interest of *individual* Bosun's dealers.

Dealer Benefits

Teak Cam Cleat Spacers make a valuable addition to the dealer's product line. They complement existing sailing accessories as well as provide the customer with the convenience of a readily available prefabricated product.

A customer who comes in to buy a cam cleat is a ready prospect for the companion spacer. Such a customer will likely want to buy mounting hardware as well.

With this unique teak product readily available, a dealer can save the customer the time and trouble of fabricating makeshift spacers.

4

■ **Figure 8–3** ■ continued

Sizable Potential Market

These Teak Cam Cleat Spacers are designed with a large and growing potential market in mind. They are custom-made for the Catalina 22, one of the most popular sailboats in use today. More than 13,000 of these sailboats have been manufactured to date. These spacers are also well suited for other similar-class sailboats.

Includes number of owners to emphasize potential sales.

Affordable Price

The Teak Cam Cleat Spacers made by Totally Teak, Inc., wholesale for $3.95/pair. Suggested retail is $6.95/pair. This low price is easy on the skipper's wallet and should help this product move well. And, of course, the obviously high profit margin should provide an incentive to your store managers.

Keeps price information short and clear.

5

■ **Figure 8–3** ■ continued

CONCLUSION

Why should a marine supply dealer consider carrying Teak Cam Cleat Spacers? This product satisfies two common criteria of sailboat owners today: it enhances the appearance of any sailboat, and it makes the boat easier to handle. The potential success of this product is based on its ability to meet these criteria and the following features and benefits:

Links list of *benefits* with their order as presented in the body of the proposal—drives home advantages of product to *user* and *dealer*.

1. It is practical, allowing quick, one-handed cleating.
2. It is ideally suited for a variety of sailors, whether they are racing, cruising, or sailing single-handedly.
3. It is a high-quality, handcrafted product that enhances the appearance of any sailboat.
4. It is a product that benefits the dealer by making a valuable addition to her or his product. It complements existing sail accessories and satisfies a customer need.
5. It is geared toward a sizable potential market. Today there are thousands of sailboats in the class for which this accessory is designed.
6. It is affordably priced and provides a good profit margin.

6

■ **Figure 8–3** ■ continued

>>> Chapter Summary

Proposals stand out as documents that aim to convince readers. You are writing to convince someone inside or outside your organization to adopt an idea, a product, or a service.

This chapter includes lists of writing guidelines for informal proposals and formal proposals. For informal proposals, follow these basic guidelines:

1. Plan well before you write.
2. Use letter or memo format.
3. Make text visually appealing.
4. Use the ABC format for organization.
5. Use the heading "Introductory Summary" for the generic abstract section.
6. Put important details in the body.
7. Use a problem/solution pattern of organization.
8. Focus attention in your conclusion.
9. Use attachments for less important details.
10. Edit carefully.

In formal proposals, abide by the same general format presented in Chapter 7 for formal reports. To be sure, formal proposals have a different tone and substance because of their more persuasive purpose. Yet they do have the same basic parts, with minor variations: cover/title page, letter/memo of transmittal, table of contents, list of illustrations, executive summary, introduction, discussion sections, conclusion, and appendices.

>>> Learning Portfolio

Collaboration at Work Proposing Changes in Security

General Instructions

Each Collaboration at Work exercise applies strategies for working in teams to chapter topics. The exercise assumes you (1) have been divided into teams of about three to six students, (2) will use team time inside or outside of class to complete the case, and (3) will produce an oral or written response. For guidelines about writing in teams, refer to Chapter 2.

Background for Assignment

Assume your school is reviewing all issues related to (1) the safety of students, faculty, and staff and (2) the security of equipment. This review has not been triggered by any particular event or problem; it is simply a periodic evaluation of conditions on campus. One step in the process has been to

request that five consulting firms visit campus, spend a day on a field investigation, and then submit a proposal that lists specific work to be done and the cost of the work. Your school will choose one firm to do the job.

Team Assignment

Assume your team is one of five consulting firms submitting a proposal. As a preliminary step, team members toured the campus, recorded observations, and collected initial ideas to propose. After discussing your observations, agree on three to five main changes to propose to the school's administration. (As an alternative, your team could focus on proposing three to five changes in one particular activity, building, or area of campus.)

Assignments

The assignments in Part 1 and Part 2 can be completed either as individual projects or as team projects. If your instructor assigns team projects, review the information in Chapter 2 on team writing.

Part 1: Short Assignments

These short assignments require either that you write parts of informal or formal proposals or that you evaluate the effectiveness of an informal proposal included here.

1. Analysis of a Request for Proposal (RFP)

Search the Internet to find an RFP, read, and analyze it. What

mytechcommlab

For practice with creating a letter of transmittal and an abstract, do the Writing a Public Library Grant Proposal Activity.

organization issued the proposal? Does the RFP identify a problem to be solved, or is it offering to fund projects through a grant? What formatting requirements does it include? Does it require qualifications such as special certifications? In small groups or as a class, compare your findings.

2. Needs Section

As this chapter suggests, informal proposals—especially those that are unsolicited—must make a special effort to establish the need for the product or service being proposed. Assume that you are writing an informal proposal to suggest a change in procedures or equipment at your

college. Keep the proposal limited to a small change; you may even see a need in the classroom where you attend class (audiovisual equipment? lighting? heating or air systems? aesthetics? soundproofing?). Write the needs section that would appear in the body of the informal proposal.

3. Revising a Conclusion

Revise the conclusion of the proposal in Figure 8–2 so that it is in paragraph form. Do you think this is an improvement over the list in the original? Write a paragraph that explains your preference.

4. Creating Boilerplate

Think about organizations to which you belong, and choose one that might submit proposals for contracts or grants. You might work for a business that could supply equipment, service, or meals to other business or government agencies, or you might be a member of a campus organization that seeks campus activity funds for special projects. Write a one paragraph description of your organization that could be used in all of its proposals.

5. Evaluation—Informal Proposal

Review the informal proposal that follows, submitted by MainAlert Security Systems. Evaluate the effectiveness of every section of the proposal.

200 Roswell Road
Marietta Georgia 30062
(770) 555–2000

September 15, 2011

Mr. Bob Montrose
Operations Manager
Lenyr Restaurant Services
16 Cuisine Way
Atlanta Georgia 30324

Dear Bob,

Thank you for giving MainAlert Security Systems an opportunity to submit a proposal for installation of an alarm system at your new office. The tour of your nearly completed office in Atlanta last week showed me all I need to know to provide you with burglary and fire protection. After reading this proposal, I think you will agree with me that my plan for your security system is perfectly suited to your needs.

This proposal describes the burglary and fire protection system I've designed for you. This proposal also describes various features of the alarm system that should be of great value. To provide you with a comprehensive description of my plan, I have assembled this proposal in five main sections:

1. Burglary Protection System
2. Fire Protection System
3. Arm/Disarm Monitoring
4. Installation Schedule
5. Installation and Monitoring Costs

BURGLARY PROTECTION SYSTEM

The burglary protection system would consist of a 46-zone MainAlert alarm control set, perimeter protection devices, and interior protection devices. The alarm system would have a strobe light and a siren to alert anyone nearby of a burglary in progress. Our system also includes a two-line dialer to alert our central station personnel of alarm and trouble conditions.

Alarm Control Set

The MainAlert alarm control set offers many features that make it well suited for your purposes. Some of these features are as follows:

1. Customer-programmable keypad codes
2. Customer-programmable entry/exit delays
3. Zone bypass option
4. Automatic reset feature
5. Point-to-point annunciation

I would like to explain the point-to-point annunciation feature, because the terminology is not as self-explanatory as the other features are. Point-to-point annunciation is a feature that enables the keypad to display the zone number of the point of protection that caused the alarm. This feature also transmits alarm-point information to our central station. Having alarm-point information available for you and the police can help prevent an unexpected confrontation with a burglar.

Interior and Perimeter Protection

The alarm system I have designed for you uses both interior and perimeter protection. For the interior protection, I plan to use motion detectors in the hallways. The perimeter protection will use glass-break detectors on the windows and door contacts on the doors.

(continued)

There are some good reasons for using both interior and perimeter protection:

1.   Interior and perimeter protection used together provide you with two lines of defense against intrusion.
2. A temporarily bypassed point of protection will not leave your office vulnerable to an undetected intrusion.
3. An employee who may be working late can still enjoy the security of the perimeter protection while leaving the interior protection off.

Although some people select only perimeter protection, it is becoming more common to add interior protection for the reasons I have given. Interior motion detection, placed at carefully selected locations, is a wise investment.

Local Alarm Signaling

The local alarm-signaling equipment consists of a 40-watt siren and a powerful strobe light. The siren and strobe will get the attention of any passerby and unnerve the most brazen burglar.

Remote Alarm Signaling

Remote alarm signaling is performed by a two-line dialer that alerts our central station to alarm and trouble conditions. The dialer uses two telephone lines so that a second line is available if one of the lines is out. Any two existing phone lines in your office can be used for the alarm system. Phone lines dedicated for alarm use are not required.

FIRE PROTECTION SYSTEM

My plan for the fire protection system includes the following equipment:

1. Ten-zone fire alarm panel
2. Eight smoke detectors
3. Water flow switch
4. Water cutoff switch
5. Four Klaxon horns

The ten-zone fire alarm panel will monitor one detection device per zone. Because each smoke detector, the water flow switch, and the water cutoff switch have a separate zone, the source of a fire alarm can be determined immediately.

To provide adequate local fire alarm signaling, this system is designed with four horns. Remote signaling for the fire alarm system is provided by the MainAlert control panel. The fire alarm would report alarm and trouble conditions to the MainAlert control panel. The MainAlert alarm control panel would, in turn, report fire alarm and fire trouble signals to our central station. The MainAlert alarm panel would not have to be set to transmit fire alarm and fire trouble signals to our central station.

ARM/DISARM MONITORING

Because 20 of your employees would have alarm codes, it is important to keep track of who enters and leaves the office outside of office hours. When an employee arms or disarms the alarm system, the alarm sends a closing or opening signal to our central station. The central station keeps a record of the employee's identity and the time the signal was received. With the arm/disarm monitoring service, our central station sends you opening/closing reports on a semi-monthly basis.

INSTALLATION SCHEDULE

Given the size of your new office, our personnel could install your alarm in three days. We could start the day after we receive approval from you. The building is now complete enough for us to start anytime. If you would prefer for the construction to be completed before we start, that would not present any problems for us. To give you an idea of how the alarm system would be laid out, I have included an attachment to this proposal showing the locations of the alarm devices.

INSTALLATION AND MONITORING COSTS

Installation and monitoring costs for your burglary and fire alarm systems as I have described them in this proposal are as follows:

- $8,200 for installation of all equipment
- $75 a month for monitoring of burglary, fire, and opening/closing signals under a two-year monitoring agreement

The $8,200 figure covers the installation of all the equipment I have mentioned in this proposal. The $75-a-month monitoring fee also includes opening/closing reports.

CONCLUSION

The MainAlert control panel, as the heart of your alarm system, is an excellent electronic security value. The MainAlert control panel is unsurpassed in its ability to report alarm status information to our central station. The perimeter and interior protection offers complete building coverage that will give you peace of mind.

The fire alarm system monitors both sprinkler flow and smoke conditions. The fire alarm system I have designed for you can provide sufficient warning to allow the fire department to save your building from catastrophic damage.

The arm/disarm reporting can help you keep track of employees who come and go outside of office hours. It is not always apparent how valuable this service can be until you need the information it can provide.

I'll call you early next week, Bob, in case you have any questions about this proposal. We will be able to start the installation as soon as you return a copy of this letter with your signature in the acceptance block.

Sincerely,
Anne Rodriguez Evans
Anne Rodriguez Evans
Commercial Sales
Enc.

ACCEPTED Lenyr Restaurant Services

By: _____

Title: _____

Date: _____

Alarm System Layout for Lenyr Restaurant Services—Atlanta, GA

Part 2: Longer Assignments

For each of these assignments (except number 9), complete a copy of the Planning Form included at the end of the book.

6. Informal Proposal

- Select a product or service (1) with which you are reasonably familiar (on the basis of your work experience, research, or other interests) and (2) that could conceivably be purchased by a local or national.
- Put yourself in the role of someone representing the company that makes the product or provides the service.
- Write an informal sales proposal in which you propose purchase of the product or service by a representative of the company.

7. Formal Proposal

Choose Option A, B, or C. Make sure that your topic is more complex than the one you would choose for the preceding informal-proposal assignments.

Option A: Community Related

- Write a formal proposal in which you propose a change in (1) the services offered by a city or town (e.g., mass transit, waste management) or (2) the structure or design of a building, garden, parking lot, shopping area, school, or other civic property.
- Select a topic that is reasonably complex and yet one about which you can locate information.
- Place yourself in the role of an outside consultant, who is proposing the change.
- Choose either an unsolicited or a solicited context.
- Write to an audience that could actually be the readers. Do enough research to identify at least two levels of audience.

Option B: School Related

- Write a proposal in which you propose a change in some feature of a school you attend or have attended.
- Choose from topics, such as operating procedures, personnel, curricula, activities, and physical plant.
- Select an audience that would actually make decisions on such a proposal.
- Give yourself the role of an outside consultant.

(?) 9. Ethics Assignment

Assume you are an employee of a local nonprofit food pantry. You have learned that a major appliance manufacturer has a grant program to provide kitchen appliances to organizations like yours. You and a volunteer, Sally, have begun working on a proposal for the program. Sally comes to you with a copy of a grant proposal that she found on the Internet—a proposal that was successfully submitted to a similar program, but sponsored by a different company. She suggests that you "re-purpose" parts of the proposal for your own proposal. After all, she argues, companies reuse boilerplate text in proposals all the time. How do you respond? Write a short essay that explains your response. Do an Internet search using terms like *proposals* and *code of ethics* to look for support for your position.

10. Informal or Formal Proposal— International Context

- Assume you are a consultant asked to propose a one-week training course to an overseas branch of an organization based in the United States. Most or all seminar participants are residents native to the country you choose—not U.S. citizens working overseas.
- Choose a seminar topic familiar to you—for example, from college courses, work experience, or hobbies—or one that you are willing to learn about quickly through some study.

- Research work habits, learning preferences, social customs, and other relevant topics concerning the country where the overseas branch is located.
- Write an informal or a formal proposal that reflects your understanding of the topic, your study of the country, and your grasp of the proposal-writing techniques presented in this chapter.

Optional Team Approach: If this assignment is done by teams within your class, assume that members of your team work for a company proposing training seminars at overseas branches around the world. Each team member has responsibility for a branch in a different country.

Different sections of the proposal will be written by different team members, who may be proposing the same seminar for all offices or different seminars. Whatever the case, the document as a whole should be unified in structure, format, and tone. It will be read by (1) the vice president for international operations at the corporate office, (2) the vice president for research and training at the corporate office, and (3) all overseas branch managers.

Chapter | 9 | Presentations

Your career will present you with many opportunities for oral presentations, both formal and informal. At the time they arise, however, you may not consider them to be "opportunities." They may seem to loom on the horizon as stressful obstacles. That response is normal. The purpose of this chapter is to provide the tools that help oral presentations contribute to your career success. The entire chapter is based on one simple principle: *Almost anyone can become an excellent speaker.* Certainly some people have more natural talent at thinking on their feet or have a more resonant voice, but success at speaking can come to all speakers, whatever their talent, if they follow the 3 Ps:

Step 1:   Prepare carefully

Step 2:   Practice often

Step 3:   Perform with enthusiasm

These steps form the foundation for all specific guidelines that follow. Before presenting these guidelines, this chapter examines specific ways that formal and informal presentations become part of your professional life.

>>> Presentations and Your Career

The following examples present some realistic situations in which the ability to speak well can lead to success for you and your organization:

- **Getting hired:** You may be asked to present information about your accomplishments to a hiring committee or small group of managers.
- **Getting customers:** If you are on a proposal team, presentations to potential clients may be a part of your regular responsibilities.
- **Keeping customers:** You may be asked to present findings or even periodic reports to clients, so that they can ask questions.
- **Contributing to your profession:** You may present new procedures that your organization has developed at professional conferences, such as the Society for Technical Communication's annual Summit.
- **Contributing to your community:** You may be asked to speak to a community group or local governing body, as a representative of your company or organization.
- **Getting promoted:** As an employee about to be considered for promotion, you may be evaluated on your ability to present information orally.

As you can see from this list, oral presentations are defined quite broadly.

Throughout your career, you will speak to different-size groups, on diverse topics, and in varied formats. The next two sections provide some common guidelines on preparation, delivery, and graphics.

>>> Guidelines for Preparation and Delivery

The goal of most presentations is quite simple: You must present a few basic points, in a fairly brief time, to an interested but usually impatient audience. If you deliver what *you* expect when *you* hear a speech, then you will give good presentations yourself.

Although the guidelines here apply to any presentation, they relate best to those that precede or follow a written report, proposal, memo, or letter. With this connection in mind, note that there are many similarities between the guidelines for good speaking and those for good writing covered in earlier chapters.

>> Presentation Guideline 1: Know Your Listeners

The following features are common to most listeners:

- They cannot "rewind the tape" of your presentation, as opposed to the way they can skip back and forth through the text of a report.
- They are impatient after the first few minutes, particularly if they do not know where a speech is going.
- They will daydream and often must have their attention brought back to the matter at hand (expect a 30-second attention span).

To respond to these realities, learn as much as possible about your listeners. For example, you can (1) consider what you already know about your audience, (2) talk with colleagues who have spoken to the same group, and (3) find out which listeners make the decisions.

Most important, make sure not to talk over anyone's head. If there are several levels of technical expertise represented by the group, decrease the technical level of your presentation accordingly. Remember—decision makers are often the ones without current technical experience. They may want only highlights; later, they can review written documents for details or solicit more technical information during the question-and-answer session after you speak.

>> Presentation Guideline 2: Use the Preacher's Maxim

The well-known preacher's maxim goes like this:

First you tell 'em what you're gonna tell 'em, then you tell 'em, and then you tell 'em what you told 'em.

This plan gives the speech a simple three-part structure that most listeners can grasp easily.

1. **Abstract (beginning of presentation):** At the outset, you should (1) get the listeners' interest (with an anecdote, a statistic, or other technique), (2) state the exact purpose of the speech, and (3) list the main points you will cover. Do not try the patience of your audience with an extended introduction—use no more than a minute.
2. **Body (middle of presentation):** Here you discuss the points mentioned briefly in the introduction, in the same order that they were mentioned. Provide the kinds of obvious transitions that help your listeners stay on track.
3. **Conclusion (end of presentation):** In the conclusion, review the main ideas covered in the body of the speech and specify actions you want to occur as a result of your presentation.

This simple three-part plan for all presentations gives listeners the handle they need to understand your speech. First, there is a clear *road map* in the introduction so that they know what lies ahead. Second, there is an organized pattern in the body, with clear transitions between points. And third, there is a strong finish that brings the audience back full circle to the main thrust of the presentation.

>> Presentation Guideline 3: Stick to a Few Main Points

Our short-term memory holds limited items. It follows that listeners are most attentive to speeches organized around a few major points. In fact, a good argument can be made for organizing information in groups of threes whenever possible. Listeners seem to remember groups of three items more than they do any other size groupings.

>> Presentation Guideline 4: Put Your Outline on Cards or Paper

The best presentations are *extemporaneous,* meaning the speaker shows great familiarity with the material but uses notes for occasional reference. Avoid extremes of reading a speech verbatim, which many listeners consider the ultimate insult, or memorizing a speech, which can make your presentation seem somewhat wooden and artificial.

Ironically, you appear more natural if you refer to notes during a presentation. Such extemporaneous speaking allows you to make last-minute changes in phrasing and emphasis that may improve delivery, rather than locking you into specific phrasing that is memorized or written out word for word.

>> Presentation Guideline 5: Practice, Practice, Practice

Many speakers prepare a well-organized speech but then fail to add the essential ingredient: practice. Constant practice distinguishes superior presentations from mediocre ones. It also helps eliminate the nervousness that most speakers feel at one time or another.

In practicing your presentation, make use of four main techniques, listed here from least to most effective:

- **Practice before a mirror:** This old-fashioned approach allows you to hear and see yourself in action. The drawback, of course, is that it is difficult to evaluate your own performance while you are speaking.

- **Use of an audio recording:** The portability of electronics allows you to practice almost anywhere. Although recording a presentation does not improve gestures, it helps you discover and eliminate verbal distractions such as *filler words* (e.g., *uhhhh, um, ya know*).

- **Use of live audience:** Groups of your colleagues, friends, or family—simulating a real audience—can provide the kinds of responses that approximate those of a real audience. Make certain that observers understand the criteria for a good presentation and are prepared to give an honest and forthright critique.

- **Use of video recording:** This practice technique allows you to see and hear yourself as others do. Your careful review of the recording can help you identify and eliminate problems with posture, eye contact, vocal patterns, and gestures.

>> Presentation Guideline 6: Speak Vigorously and Deliberately

Vigorously means with enthusiasm; *deliberately* means with care, attention, and appropriate emphasis on words and phrases. The importance of this guideline becomes clear when you think back to how you felt during the last speech you heard. At the very least, you expected the speaker to show interest in the subject and demonstrate enthusiasm.

>> Presentation Guideline 7: Avoid Filler Words

Avoiding filler words presents a tremendous challenge to most speakers. When they think about what comes next or encounter a break in the speech, they may tend to fill the gap with words and phrases, such as "um," "okay," or "you know." To eliminate such distractions, follow these three steps:

Step 1: Use pauses to your advantage. Short gaps or pauses inform the listener that you are shifting from one point to another. In signaling a transition, a pause serves to draw attention to the point you make right after the pause. Do *not* fill these strategic pauses with filler words.

Step 2: Practice with a recorder. When you play it back, you become instantly aware of fillers that occur more than once or twice. Keep a tally sheet of the fillers you use and their frequency. Your goal is to reduce this frequency with every practice session.

Step 3: Ask for help from others. After working with audio recorders in step 2, give your speech to an individual who has been instructed to stop you after each filler. This technique gives immediate reinforcement.

>> Presentation Guideline 8: Use Rhetorical Questions

Enthusiasm, of course, is your best delivery technique for capturing the attention of the audience. Another technique is the use of rhetorical questions at pivotal points in your presentation.

Rhetorical questions are those you ask to get listeners thinking about a topic, not those that you would expect them to answer out loud. They prod listeners to think about your point and set up an expectation that important information follows. Also, they break the monotony of standard declarative sentence patterns. You must make a conscious effort to insert them at points when it is most important to gain or redirect the attention of the audience. Three effective uses follow:

1. **As a grabber at the beginning of a speech:** "Have you ever wondered how you might improve the productivity of your clerical staff?"

2. **As a transition between major points:** "We've seen that a centralized copy center can improve the efficiency of report production, but will it simplify report production for your staff?"

3. **As an attention-getter right before your conclusion:** "Now that we've examined the features of a centralized copy center, what's the next step you should make at Dark Star Publishing?"

>> Presentation Guideline 9: Maintain Eye Contact

Your main goal is to keep listeners interested in what you are saying. This goal requires that you maintain control, and frequent eye contact is one good strategy.

The simple truth is that listeners pay closer attention to what you are saying when you look at them. Think how you react when a speaker makes constant eye contact with you. If you are like most people, you feel as if the speaker is speaking to you personally—even if there are 100 people in the audience. Also, you tend to feel more obligated to listen when you know that the speaker's eyes will be meeting yours throughout the presentation.

>> Presentation Guideline 10: Use Appropriate Gestures and Posture

Speaking is only one part of giving a speech; another is adopting appropriate posture and using gestures that reinforce what you are saying. Good presenters do the following:

1. Use their hands and fingers to emphasize major points.
2. Stand straight, without leaning on or gripping the lectern.
3. Step out from behind the lectern on occasion.
4. Point toward visuals on screens or charts, without losing eye contact.

With work on this facet of your presentation, you can avoid problems like keeping your hands constantly in your pockets, rustling change (remove pocket change and keys beforehand), tapping a pencil, scratching nervously, slouching over a lectern, and shifting from foot to foot.

mytechcommlab

See these guidelines at work in Model Presentation 4: Presentation to Persuade.

>>> Guidelines for Presentation Graphics

Listeners expect good graphics during oral presentations. Much like gestures, graphics transform the words of your presentation into true communication with the audience. When you display graphics and text during a presentation, they should illustrate and clarify your speech. Therefore, we include displayed text in our discussion of graphics in this section.

>> Graphics Guideline 1: Discover Listener Preferences

Some professionals prefer simple speech graphics, such as flip charts or transparencies. Others prefer more sophisticated presentations, such as animations, audio, or video. Your listeners are usually willing to indicate their preferences when you call on them. Contact the audience ahead of time and make some inquiries. Also ask for information about the room in which you will be speaking. If possible, request a setting that allows you to make best use of your graphics choice.

>> Graphics Guideline 2: Match the Graphics to the Content

Plan graphics while you prepare the text so that the final presentation seems fluid. Remember that everything you project on a screen or present on a flip chart should support and enhance your presentation.

>> Graphics Guideline 3: Keep the Message Simple

When Edward Tufte critiqued PowerPoint slides in *The Cognitive Style of PowerPoint*, one of the problems he noted was the use of too many graphic elements on each slide, the equivalent of *chartjunk* that he had argued against in his earlier studies of graphics[1] (see Chapter 3). Some basic design guidelines apply, whether you are using posters, overhead transparencies, or computer-aided graphics such as PowerPoint. Figure 9–1 is an example of an introductory slide that uses these principles.

- Use few words, emphasizing just one idea on each frame.

   Note: A common PowerPoint mistake is the use of too much text, which the speaker then reads to the audience.

- Use more white space, perhaps as much as 60%–70% per frame.

- Use landscape format more often than portrait, especially because it is the preferred default setting for most presentation software.

- Use sans-serif large print, from 14 pt. to 18 pt. minimum for text to 48 pt. for titles.

Your goal is to create graphics that are seen easily from anywhere in the room and that complement—but do not overpower—your presentation.

■ Figure 9–1 ■
Notes view of presentation slide

By avoiding asbestos contamination, you can

1. Prevent health problems
2. Satisfy regulatory requirements
3. Give yourself peace of mind

Three main reasons why building owners should be concerned about asbestos:

1. To prevent future health problems of your tenants.
2. To satisfy regulatory requirements of the government.
3. To give yourself peace of mind for the future.

---

[1]Edward R. Tufte, *The Cognitive Style of PowerPoint* (Cheshire, CT: Graphics Press, 2003).

You should also use audio and video elements sparingly. Most presentation software programs include sound effects to accompany slide changes or the appearance of text or images. These are distracting and annoying, and should be avoided. You should also use video carefully.

>> Graphics Guideline 4: Consider Alternatives to Bulleted Lists

Recently, there has been a move away from the default slide layouts in most presentation software. One recommendation is to use full-sentence headings on slides to help the audience understand and remember the information being presented; another is to combine text with graphics on slides when appropriate.[2] Figure 9–2 uses the sentence headline and image format.

>> Graphics Guideline 5: Use Colors Carefully

Colors can add flair to visuals. Use the following simple guidelines to make colors work for you:

- Have a good reason for using color (such as the need to highlight three different bars on a graph with three distinct colors).
- Be sure that a color contrasts with its background (e.g., yellow on white does not work well).
- Use no more than three or four colors in each graphic (to avoid a confused effect).

■ **Figure 9–2** ■
Notes view of presentation slide using sentence headline-image format

Prevent future health problems of your tenants.

Most important reason be concerned?

Long term health of your tenants.
Asbestos linked to:
Lung cancer
Colon cancer
Asbestosis (debilitating lung disease)

Connection first documented in 1920s, but only taken seriously in last few decades.
By this time, asbestos common in buildings.

---

[2]Michael Alley and Kathryn A. Neeley, "Rethinking the Design of Presentations Slides: A Case for Sentences Headlines and Visual Evidence." *Technical Communication 52*, no. 4 (2005): 417–426.

>> Graphics Guideline 6: Leave Graphics Up Long Enough

Because graphics reinforce text, they should be shown only while you address the particular point at hand. For example, reveal a graph just as you are saying, "As you can see from the graph, the projected revenue reaches a peak in 2007." Then pause and leave the graph up a bit longer for the audience to absorb your point.

>> Graphics Guideline 7: Avoid Handouts

Because timing is so important in your use of speech graphics, handouts are often a bad idea. Readers move through a handout at their own pace, rather than at the pace the speaker might prefer. Use them only if (1) no other visual will do, (2) your listener has requested them, or (3) you distribute them as reference material after you have finished talking.

>> Graphics Guideline 8: Maintain Eye Contact While Using Graphics

Do not stare at your visuals while you speak. Maintain control of listeners' responses by looking back and forth from the visual to faces in the audience. To point to the graphic aid, use the hand closest to the visual. Using the opposite hand causes you to cross over your torso, forcing you to turn your neck and head away from the audience.

>> Graphics Guideline 9: Include All Graphics in Your Practice Sessions

Dry runs before the actual presentation should include every graphic you plan to use, in its final form. This is a good reason to prepare graphics as you prepare text, rather than as an afterthought. If you are going to be projecting images from a transparency or computer program, the projected image may appear different than the original image. By previewing your graphics, you are able to fix them before your presentation.

You should also practice timing your graphics with your speech. Running through a final practice without graphics would be much like doing a dress rehearsal for a play without costumes and props—you would be leaving out parts that require the greatest degree of timing and orchestration.

>> Graphics Guideline 10: Plan for Technology to Fail

Murphy's Law always seems to apply when you use another person's audiovisual equipment: Whatever can go wrong, will, and at the worst possible moment. For example, a new bulb burns out and there is no extra bulb in the equipment drawer, an extension cord is too short, the screen does not stay down, the client's computer does not read your file—many speakers have experienced these problems and more. Even if the equipment works, it often operates differently from what you are used to. The only sure way to put the odds in your favor is to carry your own equipment and set it up in advance.

If you must rely on someone else's equipment, following are a few ways to prevent problems:

■ Find out exactly who is responsible for providing the equipment and contact that person in advance.

■ Have some easy-to-carry backup supplies in your car—an extension cord, a large easel pad, felt-tip markers, and dry erase markers, for example.

■ Bring handout versions of your visuals to use as a last resort.

In short, you want to avoid putting yourself in the position of having to apologize. Plan well.

>>> Chapter Summary

Anyone can become a good speaker by preparing well, practicing often, and giving an energetic performance. Your effort pays off richly by helping you deal effectively with employers, customers, and professional colleagues.

Ten guidelines for preparation and delivery lead to first-class presentations:

1. Know your listeners.
2. Use the preacher's maxim.
3. Stick to a few main points.
4. Put your outline on cards or paper.
5. Practice, practice, practice.
6. Speak vigorously and deliberately.
7. Avoid filler words.
8. Use rhetorical questions.
9. Maintain eye contact.
10. Use appropriate gestures and posture.

You should also strive to incorporate illustrations into your speeches by following these 10 guidelines:

1. Discover listener preferences.
2. Match the graphics to the content.
3. Keep the message simple.
4. Consider alternatives to bulleted lists.
5. Use colors carefully.
6. Leave graphics up long enough.
7. Avoid handouts.
8. Maintain eye contact while using graphics.
9. Include all graphics in your practice sessions.
10. Plan for technology to fail.

>>> Learning Portfolio

Collaboration at Work Speeches You Have Heard

General Instructions

Each Collaboration at Work exercise applies strategies for working in teams to chapter topics. The exercise assumes you (1) have been divided into teams of about three to six students, (2) will use team time inside or outside of class to complete the case, and (3) will produce an oral or written response. For guidelines about writing in teams, refer to Chapter 2.

Background for Assignment

Even if you have little experience as a public speaker, and even if you have not read this chapter, you already know a lot about what makes a good or bad speech because you have listened to so many presentations in your life, from informal lectures in a classroom to famous speeches by national figures. Every day you see or hear snippets of presentations in the media, so your exposure has been high. Considering all the visual input,

you probably have developed preferences for certain features in presentations.

Team Assignment

In this exercise, you and your fellow team members will do the following:

1. Share anecdotes about good and bad presentations you have heard, focusing on criteria, such as content, organization, delivery, graphics, and gestures.
2. Assemble a first list that includes features of speeches and characteristics of speakers that you consider worthy of modeling.
3. Assemble a second list that includes features of speeches and characteristics of speakers that you think should be avoided.

Create an outline and graphics for a presentation that identifies these features and explains why they are positive or negative.

Assignments

1. A 2–3 Minute Presentation Based on Your Academic Major

Give a presentation in which you discuss (a) your major field, (b) reasons for your interest in this major field, and (c) specific career paths you may pursue. Assume your audience is a group of students, with undecided majors, who may want to select your major.

2. A 5–6 Minute Presentation Based on Short Report

Select any of the short written assignments in chapters that you have already completed. Prepare a presentation based on the report you have chosen. Assume that your main objective is to present the audience with the major highlights of the written report, which they have all read. Use at least one visual aid.

3. A 5–6 Minute Presentation Based on Proposal

Prepare a presentation based on a proposal assignment at the end of Chapter 8. Assume that your audience wants highlights of your written proposal, which they have read.

5. A 10–12 Minute Presentation Based on Formal Report

Prepare a presentation based on any of the long-report assignments at the end of Chapter 7. Assume that your audience has read or skimmed the report. Your main objective is to present highlights along with some important supporting details. Use at least three visual aids.

6. Team Presentation

Prepare a team presentation in the size teams indicated by your instructor. It may be related to a collaborative writing assignment in an earlier chapter, or it may be done as a separate project. Review the Chapter 2 guidelines for collaborative work. Although related to writing, some of these suggestions apply to any team work.

mytechcommlab

Compare Presentation 2: Indian Stereotypes and Presentation 3: Status of the Organization. Which one more closely follows the Guidelines for Presentation Graphics in this chapter?

Your instructor will set time limits for the entire presentation and perhaps for individual presentations. Make sure that your team members move smoothly from one speech to the next; the individual presentations should work together for a unified effect.

7. Ethics Assignment

Suzanne Anthony, a prominent ecologist with Earhart Environmental Engineering (EEE), has been asked to make a 30-minute speech to a public workshop on environmentalism, sponsored by SprawlStopper, a regional environmental action organization. She agrees to give the talk—for which she will receive an honorarium of $500—on her area of expertise: the effects of unplanned growth on biological diversity of plant and animal species. Suzanne views the talk as a public service and has no knowledge of the sponsoring organization.

A few weeks before the speech, Suzanne's boss, Paul Finn, gets heartburn over his morning coffee as he reads an announcement about Suzanne's speech in the "Community Events" column of the local paper. Just yesterday, a large local builder, Action Homes, accepted his proposal for EEE to complete environmental site assessments on all of Action Homes' construction sites for the next three years. Paul is aware of the fact that Action Home's has an ongoing court battle with SprawlStopper concerning Action Homes' desire to develop a large site adjacent to a Civil War national park north of Atlanta. Although EEE is not now involved in the suit, Paul is worried that if Action Homes sees the name of an EEE scientist associated with an event sponsored by SprawlStopper, Action Homes may have second thoughts about having chosen EEE for its site work.

Do you think Paul should say something to Suzanne? If so, what should he say, and why should he say it? If not, why not? If you were in Suzanne's place, how would you respond to a suggestion by Paul that her speech might be inappropriate? Are there any similarities between the situation described here and the one characterized in this chapter's "Communication Challenge"?

8. International Communication Assignment

Prepare a team presentation which results from research your team does on the Internet concerning speech communication in a country outside the United States.

Option A: Retrieve information about one or more businesses or careers in a particular country. Once you have split up the team's initial tasks, conduct some of your business by e-mail, and then present the results of your investigation in a panel presentation to the class. For example, your topic could be the computer software industry in England, the tourist industry in Costa Rica, or the textile industry in Malaysia.

Option B: Retrieve information on subjects related to this chapter—for example, features of public speaking, business presentations, presentation graphics, and meeting management.

Chapter | **10** | The Job Search

In applying for a job, you must assess your abilities, find an appropriate job match, and persuade a potential employer that you are the right one for the job.

This chapter offers suggestions on these main activities:

- Researching occupations and companies
- Writing job letters and resumes
- Succeeding in job interviews

>>> Researching Occupations and Companies

Before writing a job letter and resume, you will need information about (1) career fields that interest you, (2) specific companies that hire graduates in your field, and, (3) specific jobs that are available. Following are some pointers for finding such information.

>> Do Basic Research in Your College Library or Placement Office

Libraries and placement centers offer one starting point for getting information about professions. Following are a few well-known handbooks and bibliographies found in reference collections. They either give information about occupations or provide names of other books that supply such information:

Career Choices Encyclopedia: Guide to Entry-Level Jobs

Directory of Career Training and Development Programs

High-Technology Careers

Occupational Outlook Handbook

Professional Careers Sourcebook: An Information Guide for Career Planning

You can also check online sources, such as www.careeroverview.com.

>> Interview Someone in Your Field of Interest

To get the most current information, arrange an interview with someone working in an occupation that interests you. This source often goes untapped by college students, who mistakenly think such interviews are difficult to arrange. In fact, you can usually locate people to interview through (1) your college placement office, (2) your college alumni association, or (3) your own network of family and friends. Another possibility is to call a reputable firm in the field and explain that you wish to interview someone in a certain occupation.

Once you set up the interview, prepare well by listing your questions in a notebook or on a clipboard that you take with you to the interview. This preparation keeps you on track and shows those being interviewed that you value their time and information. Following are some questions to ask:

- How did you prepare for the career or position you now have?
- What college course work or other training was most useful?
- What types of activities fill your typical working day?
- What features of your career do you like the most? What features of your career do you like the least?
- What personality characteristics are most useful to someone in your career?
- How would you describe the long-term outlook of your field?
- Do you know any books, periodicals, or online sources that might help me find out more about your field?
- Do you know any individuals who, like you, might permit themselves to be interviewed about their choice of a profession?

Although this interview may lead to a discussion about job openings in the interviewer's organization, the main purpose of the conversation is to retrieve information about an occupation.

>> Find Information on Companies in Your Field

With your focus on a profession, you can begin screening companies that employ people in the field chosen. Determine the types of information you want to find; examples include location, net worth, number of employees, number of workers in your specific field, number of divisions, types of products or services, financial rating, and names and titles of company officers. The following are some sources that might include such information. They can be found in the reference sections of libraries and in college placement centers.

The Career Guide gives overviews of many American companies and includes information, such as types of employees hired, training opportunities, and fringe benefits.

Corporate Technology Directory profiles high-tech firms and covers topics, such as sales figures, number of employees, locations, and names of executives.

Facts on File Directory of Major Public Corporations gives essential information on 5,700 of the largest U.S. companies listed on major stock exchanges.

Job Choices in Science and Engineering, an annual magazine published by the College Placement Council, includes helpful articles and information about hundreds of companies that hire technical graduates.

Peterson's Business and Management Jobs provides background on employers of business, management, and liberal-arts graduates.

Peterson's Engineering, Science, and Computer Jobs provides background information on employers of technical graduates.

Standard & Poor's Register of Corporations, Directors, and Executives lists names and titles of officials at 55,000 public and private U.S. corporations.

>> Do Intensive Research on a Selected List of Potential Employers

The previous steps get you started finding information on occupations and firms. Ultimately, you will develop a selected list of firms that interest you. Your research may have led you to these companies, or your college placement office may have told you that there are job openings. Now you must conduct an intensive search to learn as much as you can about the firms. Following are a few sources of information, along with the kinds of questions each source helps answer:

- **Annual reports** (often available at company Web sites and in your library or placement office): How does the firm describe its year's activities to stockholders? What are its products or services?

- **Web sites or media kits** (available online or from public relations offices): How does the firm portray itself to the public? Is this firm "Internet-savvy"? What can you infer about the firm's corporate culture?

- **Personnel manuals and other policy guidelines:** What are features of the firm's *corporate culture*? How committed is the firm to training? What are the benefits and retirement programs? Where are its branches? What are its customary career paths?

- **Graduates of your college or university now working for the firm:** What sort of reputation does your school have among decision makers at the firm?

- **Company newsletters and in-house magazines:** How open and informative is the firm's internal communication?

- **Business sections of newspapers and magazines:** What kind of news gets generated about the firm?

- **Professional organizations or associations:** Is the firm active within its profession?

- **Stock reports:** Is the firm making money? How has it done in the past five years?

- **Accrediting agencies or organizations:** How has the firm fared during peer evaluations?

- **Former employees of the company:** Why have people left the firm?

- **Current employees of the company:** What do employees like, or dislike, about the company? Why do they stay?

In other words, you should thoroughly examine an organization from the outside. The information you gather helps you decide where to apply and, if you later receive a job offer, where to begin or continue your career.

>> Use Your Computer to Gather Data

Today, many applicants go directly to the computer to find information about professions, organizations, graduate schools, and job openings. No doubt between the time this book is written and then published, the names and number of online resources will change dramatically. Generally, some of the information available includes the following:

- College and university catalogs
- Web sites for companies, organizations, and schools
- Employment listings from local and national sources
- Online discussion forums involving recent graduates of colleges and universities

There are a number of job search sites available on the Internet. At many of these sites, you can

1. Get tips on writing your resume and searching for jobs
2. Post your resume at the site, in multiple forms for varied employers
3. Peruse lists of job openings
4. Receive advice from professionals who are in careers you wish to pursue
5. Read in-depth reports on particular companies
6. Share comments with other job seekers

In other words, the Internet helps you locate a variety of information during your job search. Moreover, you can use your computer to search for openings and respond to job advertisements, as mentioned in the next section.

mytechcommlab

Career Letter 3: Application shows that the writer has researched the company to which she is applying.

>>> Job Correspondence

Job letters and resumes must grab the attention of busy readers, who may spend less than 60 seconds deciding whether to consider you further. This section gives you the tools to write a successful letter and resume. *Successful,* of course, means a letter and resume that get you an interview.

Many job letters and resumes still get sent through the mail. However, a growing number of companies accept applications through the Internet. For example, online services place resumes into a database used by hundreds and perhaps thousands of companies. The resumes are scanned with software that searches for key words that reflect abilities needed for specific jobs. The program then sends selected resumes to companies. If you use this kind of service, do not expect the level of confidentiality and security that you have with personal mail.

Whether you use online techniques like e-mail and resume services or stick with the traditional approach, the same basic writing guidelines apply. Your letter, usually no longer than one page, should be specific about the job you seek and your main selling points. Then the resume—usually no more than one or two pages—should highlight your background.

Job Letters

A job letter is just another type of sales letter—except that you are selling yourself, not a product or service (see Figure 10–1). In preparing to write one, take the point of view of the persons to whom you are writing. What criteria do they use to evaluate your credentials?

1523 River Lane
Worthville OH 43804
April 6, 2011

Mr Willard Yancy
Director
Automotive Systems
XYZ Motor Company
Product Development Division
Charlotte NC 28202

Dear Mr. Yancy:

Recently I have been researching the leading national companies in automotive computer systems. Your position announcement in the April 6 *National Business Employment Weekly* caught my eye because of XYZ's innovations in computer-controlled safety systems. I would like to apply for the automotive computer engineer opening.

Your advertisement notes that experience in computer systems for machinery or robotic systems would be a plus. I have had extensive experience in the military with computer systems, ranging from a digital communications computer to an air traffic control training simulator. In addition, my college experience includes courses in computer engineering that have broadened my experience. I am eager to apply what I have learned to your company.

My mechanical knowledge was gained from growing up on my family's dairy farm. After watching and learning from my father, I learned to repair internal combustion engines, diesel engines, and hydraulic systems. Then for five years I managed the entire dairy operation.

With my training and hands-on experience, I believe I can contribute to your company. Please contact me at 614/555-2731 if you wish to arrange an interview.
Sincerely,

James M. Sistrunk

James M. Sistrunk

Enclosure: Resume

■ **Figure 10–1** ■ Job application letter (block style)

How much or how little do they want in the letter? What main points are they hunting for as they scan your resume? This section examines the needs of these readers and gives guidelines for you, the writer.

The Readers' Needs

You probably will not know personally the readers of your job letter, so you must think hard about what they may want. Your task is complicated by the fact that often there are several readers of your letter and resume who may have quite different backgrounds.

One possible scenario follows:

Step 1:   The letter may go first to the personnel office, where a staff member specializing in employment selects letters and resumes that meet the criteria stated in the position announcement. (In some large employers, letters and resumes may even be stored in a computer, where they are scanned for key words that relate to specific jobs.)

Step 2:   Applications that pass this screening are sent to the department manager who supervises the employee that is hired. The manager may then select a group to be interviewed. This manager interviews applicants and ultimately hires the employee.

One variation of this process has the human resources department doing an interview as well as screening letters and resumes—before the department manager even hears about any applications. Another variation, as noted earlier, has the employer relying on an online resume service for the initial screening.

Yet sooner or later, a supervisor or manager reads your letter and resume. And most readers, whatever their professional background, have the following five characteristics in common:

>> Feature 1: They Read Job Letters in Stacks

Most search-and-screen processes are such that letters get filed until there are many to evaluate. Your reader faces this intimidating pile of paper.

>> Feature 2: They Are Tired

Some employment specialists save job letters for later reading, but many people who do hiring get to job letters at the end of a busy day, so they have even less patience than usual for flowery wording or hard-to-read typefaces.

>> Feature 3: They Are Impatient

Your readers expect major points to jump right out at them. Usually they will not dig for information that cannot be found quickly.

>> Feature 4: They Become Picky Grammarians

Readers of all backgrounds expect good writing when they read job letters. There is an assumption that a letter asking for a job should reflect solid use of the language. If the letter does contain a typo or grammar error, the reader may wonder about the quality of writing you will produce on the job.

>> Feature 5: They Want Attention Grabbers but Not Slickness

You want the content of your letter and resume to stand out without the use of gimmicks. For example, white or off-white stationery is still the standard, along with traditional fonts with lots of white space for easy reading.

Of course, likes and dislikes vary. An advertising director, who works all day with graphics, may want a bolder format design than an engineering manager, who works with documents that are less flashy. If you cannot decide, it is best to use a conservative format and style.

The Letter's Organization

Your one goal is to tantalize the reader enough to want to interview you—that is all. With that goal and the reader's needs in mind, your job letter should follow the ABC format on the left.

This pattern gives you a starting point, but it is not the whole story. There is one feature of application letters that cannot be placed easily in a formula. Work hard with your draft to develop a unity and flow that, by itself, sets you apart from the crowd. Your attention grabber engages interest, but the clarity of your prose persuades readers that you are an applicant to be interviewed.

mytechcommlab

Career Letter 5: Inquiry displays effective strategies for an unsolicited application.

ABC Format: Job Letters

- **ABSTRACT:** Apply for a specific job
  - Refer to advertisement, mutual friend, or other source of information about the job
  - Briefly state what makes you and outstanding candidate who can uniquely meet the main need of your potential employer
- **BODY:** Specify your understanding of the reader's main needs
  - Provide main qualifications that satisfy these needs (but only highlight points from the resume—do not simply repeat all of the resume information)
  - Address specific qualifications mentioned in a job announcement
  - Avoid mentioning weak points or deficiencies
  - Keep body paragraphs to six or fewer lines
  - Use a bulleted or numbered list if it helps draw attention to three or four main points
  - Maintain the "you" attitude throughout
- **CONCLUSION:** Tie the letter together with one main theme or selling point, as you would a sales letter
  - Refer to your resume
  - Explain how and when the reader can contact you for an interview

Resumes

Resumes usually accompany application letters. Three points make writing resumes a challenge:

1. **Emphasis:** You should select just a *few major points of emphasis* from your academic and professional life.

2. **Length:** You often should use only *one page*. For individuals with extensive experience, a two-page or more resume is acceptable.

3. **Arrangement:** You should arrange information so that it is *pleasing to the eye and easy to scan*. Prospective employers spend less than a minute assessing your application. They may even use computers to scan resumes, taking even less time.

Computers pose a special challenge to a resume writer because they fail to appreciate some of the elegant variations sometimes used to get attention. If you are writing a resume that may be read by a computer, you may want to (1) use white or very light-colored paper; (2) focus on key words—especially job skills—that might be picked up by the computer scan; and (3) avoid design features that might present obstacles to the scan, such as italics, fancy typefaces, and graphics. You may also need to format a resume that can be copied and pasted into forms for online applications. Such resumes should be saved as text files (.txt). It is a good idea to create this kind of resume in Notepad or a similar program on your computer. Avoid using tabs, italics, bolding, bullets, or other special characters, as these do not translate to the text file. (For an example of a resume formatted to be submitted online, see Figure 10–2.)

This section distills the best qualities of many formats into two basic patterns:

1. The chronological resume, which emphasizes employment history.
2. The functional resume, which emphasizes the skills you have developed.

The following paragraphs describe the main parts of the resume. The "Experience" section explains the differences between chronological, functional, and combined resumes.

Objective

Human resource directors, other people in the employment cycle, and even computers may sort resumes by the Objective statement. Writing a good objective is hard work, especially for new graduates, who often just want a chance to start working at a firm at any level. Do not make the mistake of writing an all-encompassing statement such as: "Seeking challenging position in innovative firm in civil-engineering field." It gives the impression that you have not set clear professional goals.

Most objectives should be one sentence. They should be detailed enough to show that you are interested in a specific career, yet open-ended enough to reflect a degree of flexibility. If you have several quite different career options, you might want to design a different resume for each job description, rather than trying to write a job objective that takes in too much territory.

Education

Whether you follow the objective with the "Education" or "Experience" section depends on the answer to one question: Which topic is most important to the reader? Most recent college graduates lead off with "Education," particularly if the completion of the degree prompted the job search.

Obligatory information includes your school, school location, degree, and date of graduation. It is what you include beyond the bare details, however, that most interests employers. Following are some possibilities:

■ **Grade point average:** Include it if you are proud of it; do not if it fails to help your case.

■ **Honors:** List anything that sets you apart from the crowd—such as dean's list or individual awards in your major department. If you have many, include a separate "Recognitions" heading toward the end of the resume.

```
JAMES M SISTRUNK

1523 River Lane
Worthville OH 43804
(614) 555-2731

EMAIL
jmsistrunk@tmail.com

OBJECTIVE
To contribute to the research, design, and development of
automotive computer control systems

EDUCATION
Bachelor of Science in Computer Engineering
Northern College of Technology
Shipley
PA June 2011
3.5 GPA (out of 4.0 scale)

Computer Repair Technician Certification Training
U.S. Air Force Technical Training Center
Keesler Air Force Base
Biloxi
MS 2002-2003

SKILLS AND EXPERIENCE
Leadership
* Supervised processing and orientation for new students from
basic training
* Responsible for dairy operations on 500-acre farm Computers
and software
* Expert in C++ and Object Oriented Languages
* Three years experience in diagnostics and troubleshooting of
mainframe computer systems

Languages
* Fluent in German

ACTIVITIES AND HONORS
Association of Computing Machinery (ACM)
Dean's List
8 quarters
Award of an Air Force Specialty Code "5" skill level
Coached middle school teams in U.S. RobotOlympics

EMPLOYMENT HISTORY
2002-2007 Computer Repair Technician U.S. Air Force
1997-2002 Assistant Manager Spring Farm Wootan
Ohio

REFERENCES
Available on request
```

■ **Figure 10–2** ■ Resume formatted for online submission

- **Minors:** Highlight any minors or degree options, whether they are inside or outside your major field. Employers place value on this specialized training, even if (and sometimes especially if) it is outside your major field.

- **Key courses:** When there is room, provide a short list of courses you consider most appropriate for the kind of position you are seeking. Because the employer probably will not look at your transcripts until a later stage of the hiring process, use this brief listing as an attention grabber.

Activities, Recognitions, Interests

Most resumes use one or two of these headings to provide the reader with additional background information. The choice of which, if any, to use depends on which you think best support your job objective. Following are some possibilities:

- **Activities:** Selected items that show your involvement in your college or your community or both.

- **Recognitions:** Awards and other specific honors that set you apart from other applicants. (Do not include awards that might appear obscure, meaningless, or dated to the reader, such as most high-school honors.)

- **Interests:** Hobbies or other interests that give the reader a brief look at the "other" you.

However you handle these sections, they should be brief and not detract from the longer and more significant sections described previously.

Experience

This section poses a problem for many applicants just graduating from college. Depending on the amount of work experience you have gained, consider three options for completing this section of the resume: (1) emphasize specific positions you have held (chronological resume), (2) emphasize specific skills you have developed in your experience (functional resume), or (3) emphasize both experience and skills (combined resume).

>> Option 1: Chronological Format

This option works best if your job experience has led logically toward the job you now seek. Figure 10–3 is an example of a chronological resume that meets these guidelines:

- List relevant full-time or part-time experience, including internships, in reverse chronological order.
- Be specific about your job responsibilities while still being brief.
- Be selective if you have had more jobs than can fit on a one-page resume.

James M. Sistrunk
1523 River Lane
Worthville OH 43804
(614) 555-2731
jmsistrunk@tmail.com

OBJECTIVE: To contribute to the research, design, and development of automotive computer control systems

EDUCATION: Bachelor of Science in Computer Engineering
(expected June 2011)
      Northern College of Technology
      Shipley
      PA
      3.5 GPA (out of 4.0 scale)
Computer Repair Technician Certification Training, 2002–2003
      U.S. Air Force Technical Training Center
      Keesler Air Force Base
      Biloxi
      MS

Major Courses:

Semiconductor Circuits & Devices	Artificial Machine Intelligence
Robotic Systems	Communication Control Systems
Microprocessor Control	Microcomputer Applications
Microcomputer Systems	Digital Control Systems

Related Courses:

C++	Programming Languages
Business Communication	Engineering Economy
Industrial Psychology	Technical Communication
Other Skills:	Fluent in German

ACTIVITIES AND   Association of Computing Machinery (ACM)
HONORS: Coached middle school teams in U.S. RobotOlympics
Dean's List
8 quarters
U.S. Air Force Secret Clearance
Award of an Air Force Specialty Code "5" skill level

EMPLOYMENT:
2002–2007   Computer Repair Technician
     U.S. Air Force
1997–2002   Assistant Manager
     Spring Farm
     Wootan
     Ohio

REFERENCES: Available upon request

■ **Figure 10–3** ■ Chronological resume emphasizing education

- Include nonprofessional tasks (such as working on the campus custodial staff) if it helps your case (e.g., the employer might want to know that you worked your way through college).

- Remember that if you leave out some jobs, the interview will give you the chance to elaborate on your work experience.

- Select a readable format with appropriate white space.

- Use action verbs and lists to emphasize what you did or what you learned at jobs—for example, "Provided telephone support to users of System/23." Use parallel form in each list.

> mytechcommlab
>
> Model Resume 2 is another example of a chronological resume.

>> **Option 2:** Functional Format

This approach works best if (1) you wish to emphasize the skills and strengths you have developed in your career rather than specific jobs you have had or (2) you have had "gaps" in your work history that would be obvious if you used the chronological format. Sometimes your skills built up over time may be the best argument for your being considered for a position, even if your job experience also is strong. For example, you may have five years' experience in responsible positions at four different retailers. You then decide to write a functional resume focusing on the three skill areas you developed: sales, inventory control, and management. Figure 10–4 is an example of a functional resume.

If you write a functional resume that stresses skills, you may want to follow this section with a brief employment history. Most employers want to know where and when you worked, even though this issue is not a high priority. *Note:* If you decide to leave out the history, bring it with you to the interview on a separate sheet.

> mytechcommlab
>
> Model Resume 1 is another example of a functional resume.

References

Your resume opens the door to the job interview and later stages of the job process, when references will be called. There are two main approaches to the reference section of the resume:

1. Writing "Available upon request" at the end of the page
2. Listing names, addresses, and phone numbers at the end of the resume or on a separate page

The first assumes the reader prefers contacting you before references are solicited. The second assumes the reader prefers to call or write references directly, without having to contact you first. Use the format most likely to meet the needs of a particular employer.

Your goal is to write an honest resume that emphasizes your good points and minimizes your deficiencies. To repeat a point made at the outset, you want your resume and job letter to open the door for later stages of the application process. Indeed, they represent a personal sales letter, for what you are selling is the potential you offer to change an organization and, perhaps, the world as well. Considering such heady possibilities, make sure to spend the time necessary to produce first-rate results.

James M. Sistrunk
1523 River Lane
Worthville OH 43804
(614) 555-2731
jmsistrunk@tmail.com

Professional Objective:

To contribute to the research, design, and development of automotive computer control systems

Education:

B.S., Computer Engineering, (expected June 2011)
    Northern College of Technology
    Shipley
    PA
    3.5 GPA (out of 4.0 scale)
Computer Repair Technician Certification Training, 2002–2003
    U.S. Air Force Technical Training Center
    Keesler Air Force Base
    Biloxi
    MS

Certifications and Awards:

Dean's List
8 quarters
Computer Repair Technician Certification Training
U.S. Air Force Secret Clearance
Award of an Air Force Specialty Code "5" skill level

Skills and Experience:

Management	Responsible for dairy operations on 500-acre farm
Training	Supervised processing and orientation for new students from basic training
Computers and software	Expert in C++ and Object Oriented Languages Three years experience in diagnostics and troubleshooting of mainframe computer systems
Robotics	Coached middle school teams in U.S. Robot-Olympics. Team placed Second in 2008.
Mechanical skills	Conducted preventative maintenance inspections Repaired sophisticated farm equipment
Other skills	Fluent in German

References:

Available upon request

■ **Figure 10–4** ■ Functional resume emphasizing skills and experience

>>> Job Interviews

Your job letter and resume have only one purpose: to secure a personal interview. Much has been written about job interviews. Most of the good advice about interviewing goes back to just plain common sense about dealing with people. Following are some suggestions to show you how to prepare for a job interview, perform at your best, and send a follow-up letter.

Preparation

>> Do Your Homework on the Organization

Once you have been selected for an interview, review whatever information you have already gathered about the employer. Then go one step further by searching for current information. Your last source may be someone you know at the organization, or a friend of a friend.

When you do not have personal contacts, use your research skills again. For large firms, locate recent periodical or newspaper articles by consulting general indexes—such as the *Business Periodicals Index, Wall Street Journal Index, Readers' Guide to Periodicals, New York Times Index*, or the index for any newspaper in a large metropolitan area. For smaller firms, consult recent issues of local newspapers for announcements about the company. As noted earlier, the company's Web site can also be a good source of current information about an organization.

>> Write Out Answers to the Questions You Consider Likely

Although you obviously would not take written answers with you to the interview, writing them out for your review ahead of time will give you a level of confidence unmatched by candidates who only ponder questions that might come their way.

There are few original questions asked in job interviews. Most interviewers simply select from some standard questions to help them find out more about you and your background. Following are some typical questions, along with tips for responses:

1. **Tell me a little about yourself.** Keep your answer brief and relate it to the position and company—do not wander off into unrelated issues, like hobbies, unless asked to do so.

2. **Why did you choose your college or university?** Be sure your main reasons relate to academics—for example, the academic standing of the department, the reputation of the faculty, or the job placement statistics in your field.

3. **What are your strengths?** Focus on two or three qualities that would directly or indirectly lead to success in the position for which you are applying.

4. **What are your weaknesses?** Choose weaknesses that if viewed from another perspective, could be considered strengths—for example, your perfectionism or overattention to detail.

5. **Why do you think you would fit into this company?** Using your research on the firm, cite several points about the company that correspond to your own professional interests—for example, the firm may offer services in three fields that relate to your academic or work experience.

6. **What jobs have you held?** Use this question as a way to show that each previous position, no matter how modest, has helped prepare you for this position—for example, part-time employment in a fast-food restaurant developed teamwork and interpersonal skills.

7. **What are your long-term goals?** Be ready to give a 5- or 10-year plan that, preferably, fits within the corporate goals and structure of the firm to which you are applying—for example, you may want to move from the position of technical field engineer into the role of a project manager, developing your management skills.

8. **What salary range are you considering?** Avoid discussing salary if you can. Instead, note that you are most interested in criteria, such as job satisfaction and professional growth. If pushed, give a salary range that is in line with the research you did on the career field in general and the company in particular; see the section in this chapter regarding negotiating.

9. **Do you like working in teams or prefer working alone?** Most employers want to know that you have interest and experience in teamwork—whether in college courses or previous jobs, but they also admire and reward individual accomplishment. In deciding what part of your background to emphasize, consider the corporate culture of the organization interviewing you.

10. **Do you have any questions for me?** Always be ready with questions that reinforce your interest in the organization and your knowledge of the position—for example, "Given the recent opening of your Tucson warehouse, do you plan other expansions in the Southwest?" or "What types of in-house or off-site training do you offer new engineers who are moving toward project management?" Other questions can concern issues, such as (a) benefits, (b) promotions, (c) type of computer network, and (d) travel requirements.

>> Do Mock Interviews

You improve your chances by practicing for job interviews. One of the best techniques is role-playing. Ask a friend to serve as the interviewer and give him or her a list of questions from which to choose. Also, inform that person about the company so that he or she can improvise during the session. This way, you are prepared for the real thing.

You can get additional feedback about your interviewing abilities by videotaping your role-playing session. Reviewing the videotape helps you highlight questions that pose special problems for you and mannerisms that need correction. This technique is especially useful if you are one of the growing number of applicants who take part in a video interview with a recruiter.

>> Be Physically Prepared for the Interview

Like oral presentations, job interviews work best when you are physically at your best:

- Get a good night's rest before the interview.
- Avoid caffeine or other stimulants.
- Eat about an hour beforehand so that you are not distracted by hunger pangs during the session.
- Take a brisk walk to dispel nervous energy.

Performance

Good planning is your best assurance of a successful interview. Of course, there are always surprises that may catch you. Remember, however, that most interviewers are seriously interested in your application and want you to succeed. Help them by selling yourself and thus giving them a reason to hire you. Following are some guidelines for the interview.

>> Dress Appropriately

Some practical suggestions for interview attire include the following:

- Dress conservatively and avoid drawing attention to your clothing—for example, do not use the interview to break in a garment in the newest style.
- Consider the organization—for example, a brokerage-firm interview may require a dark suit for a man and a tailored suit for a woman, whereas other interviews may require less formal attire.
- Avoid excessive jewelry.
- Pay attention to the fine points—for example, wear shined shoes and carry a tasteful briefcase or notebook.

>> Take an Assertive Approach

Be positive, direct, and unflappable. Use every question as a springboard to show your capabilities and interest. To be sure, the degree to which you assert yourself partly depends on your interpretations of the interviewer's preference and style. Although you do not want to appear pushy, you should take the right opportunities to sell yourself and your abilities.

>> Use the First Few Minutes to Set the Tone

What you have heard about first impressions is true: Interviewers draw conclusions quickly. Having given many interviews, they are looking for an applicant who injects vitality into the interview. Within a minute or two, establish the themes and the tone that will be reinforced throughout the conversation. In this sense, the interview subscribes to the Preacher's Maxim mentioned in Chapter 9: "First you tell 'em what you're gonna tell 'em, then you tell 'em, and then you tell 'em what you told 'em."

>> Maintain Eye Contact While You Speak

Although you may want to look away occasionally, much of the time your eyes should remain fixed on the person interviewing you. If you are being interviewed by several people, make eye contact with all of them throughout the interview. You are never quite certain exactly who may be the decision maker in your case.

>> Be Specific in the Body of the Interview

In every question, find the opportunity to say something specific about you and your background. For example, rather than simply stating that your degree program in computer science prepared you for the open position, cite three specific courses and summarize their relevance to the job.

>> Do Not Hesitate

A job interview is no time to hesitate, unless you are convinced the job is not for you. The question is this: Do you want the job or not? If you do, then accept the requirements of the position and show excitement about the possibilities. You can always turn down the job if you receive an offer and decide later that some restrictions, like travel, are too demanding.

>> Reinforce Main Points

Orchestrate the end of the interview so that you have the chance to summarize your interest in the position and your qualifications. Here is your chance to follow through on the "Tell 'em what you told 'em" part of the Preacher's Maxim.

Follow-Up Letters

Follow every personal contact with a letter or e-mail to the person with whom you spoke. Send it within 24 hours so that it immediately reinforces the person's recollection of you. This simple strategy gives you a powerful tool for showing interest in a job. Follow-up letters abide by the same basic letter pattern discussed in Chapter 4.

- Write no more than one page.
- Use a short first paragraph to express appreciation for the interview.
- Use the middle paragraph(s) to (a) reinforce a few reasons why you would be the right choice for the position or (b) express interest in something specific about the organization.
- Use a short last paragraph to restate your interest in the job and to provide a hopeful closing.

See Chapter 4 for the various formats appropriate for all types of business letters.

When your audience might appreciate a less formal response, consider writing your interviewer a personal note instead of a typed letter. This sort of note is most appropriate when you plan a short message.

mytechcommlab

Career Letter 7 is an example of a thank you note that is sent after an interview.

>>> Chapter Summary

This chapter surveys the entire process of searching for jobs. As a first step in the process, learn about occupations and specific employers that interest you. Second, write letters and resumes that get attention and respond to specific needs of employers. You can choose from chronological, functional, or combined resume formats, using the patterns of organization and style that best highlight your background. Third, prepare carefully for your job interview, especially in anticipating the questions that may be asked. Then perform with confidence. Also, do not forget to send a thank-you letter or e-mail soon after the interview.

>>> Learning Portfolio

Collaboration at Work　Planning for Success

General Instructions

Each Collaboration at Work exercise applies strategies for working in teams to chapter topics. The exercise assumes you (1) have been divided into teams of about three to six students, (2) will use team time inside or outside of class to complete the case, and (3) will produce an oral or written response. For guidelines about writing in teams, refer to Chapter 2.

Background for Assignment

Some students attend college for its own sake because they love learning; others may like to learn but mainly view college as a stepping-stone to a career. Because most students are in the second group, they give a good deal of thought to what they will do with their working lives. If they do not, they should. Achieving success in a career starts with establishing a careful plan for getting from here to there. It is a process that involves considerably more than gaining a degree.

Team Assignment

For this exercise, you and your team members must brainstorm about a strategy for getting a particular job or entering a particular career desired by one of your team's members. (You can do the exercise for several careers, if you have time.) Following are the three steps:

1. Share information about the ideal jobs of all team members.
2. Choose one position or career that best fits the assignment (i.e., one that lends itself to being achieved in an incremental process that is doable).
3. Generate a list of specific steps for getting the position or entering the career (include deadlines and criteria for success at each stage).

If you have time to conduct some research and have access to the Internet, go to one of the Web sites listed on page 204 in this chapter for assistance in your research.

Assignments

mytechcommlab

For more practice with chronological resumes, do the Revising a Chronological Resume Activity.

1. Chronological Resume

Using the guidelines on pages 211–213, create a chronological resume which you could use to apply for an internship or full time position.

mytechcommlab

For more practice with functional resumes, do the Writing a Functional Resume Activity.

2. Functional Resume

Using the guidelines on pages 213–214, create a functional resume which you could use to apply for an internship or full time position.

3. Resume Formatted for Online Submission

Start with the resume you created for Assignment 1 or Assignment 2, and format it for submission online, using the guidelines on page 204.

4. Research on a potential employer

Identify an organization to which you could submit a job application. Use as many of the resources listed on pages 213–214 to answer these questions:

- How does the firm describe its year's activities to stockholders? What are its products or services?
- How does the firm portray itself to the public? Is this firm "Internet-savvy"? What can you infer about the firm's corporate culture?
- What are features of the firm's *corporate culture*? How committed is the firm to training? What are the benefits and retirement programs? Where are its branches? What are its customary career paths?
- What sort of reputation does your school have among decision makers at the firm?
- How open and informative is the firm's internal communication?
- What kind of news gets generated about the firm?
- Is the firm active within its profession?
- Is the firm making money? How has it done in the past five years?
- How has the firm fared during peer evaluations?
- Why have people left the firm?
- What do employees like, or dislike, about the company? Why do they stay?

5. Job Letter

Find a job advertisement in the newspaper, on the Internet, or at your college placement office. The advertisement should match either qualifications you have now or those you plan to have after you complete the academic program on which you are now working. Write a job letter that responds to the advertisement. Submit the letter and written advertisement to your instructor.

If useful for this assignment and if permitted by your instructor, you may fictionalize part of your resume so that it lists a completed degree program and other experience not yet acquired. This way, the letter and resume reflect the background you would have if you were applying for the job. Choose the resume format that best fits your credentials.

As an alternative, write a letter and resume to apply for an internship in your major field. To find out about internships, contact your department or campus internship director, or ask about internships at your college placement office.

6. Job Interview

Pair up with another classmate for this assignment. First, exchange the letters, resumes, and job advertisements referred to in Assignment 1. Discuss the job advertisements so that you are familiar with the job being sought by your counterpart, and vice versa. Then perform a role-playing exercise during which you act out the two interviews, one person as applicant and the other as interviewer.

Option: Include a third member in your team. Have this person serve as a recorder, providing an oral critique of each interview at the end of the exercise. Then collaborate among the three of you in producing a written critique of the role-playing exercise. Specifically, explain what the exercise taught you about the main challenges of the job interview.

7. Follow-Up Thank You Letter

Write a follow-up letter to the interview that resulted from Assignment 2.

8. Ethics Assignment

Searching for employment presents job seekers with some ethical challenges. A few "ethically challenged" individuals paint the portrait in their resume of someone who only remotely resembles the real thing. Certainly, lying and deception occur, but most writers simply want to present what they have accomplished and learned in the best possible light. As this chapter suggests, the resume is no time

to be overly modest. With the goal of supportable self-promotion in mind, evaluate the degree to which the following resume entries are accurate representations of the facts that follow them.

1. **Resume Entry:** June–September 2008—Served as apprentice reporter for a Detroit area weekly newspaper.
Reality: Worked for a little over three months as a fact-checker for a group of reporters. Was let go when the assistant editor decided to offer the apprentice position to another, more-promising individual with more journalistic experience.

2. **Resume Entry:** July 2007—Participated in university-sponsored trip to Germany.
Reality: Flew to Germany with two fraternity brothers for a two-day fraternity convention in Munich, after which the three of you toured Bavaria for a week in a rental car.

3. **Resume Entry:** Summer 2007, 2008, 2009—Worked for Berea Pharmacy as a stock clerk, salesperson, and accountant.
Reality: Helped off and on with the family business, Berea Pharmacy, during three summers while in college—placing merchandise on shelves, working on the cash register, and tallying sales at the end of the day. Your father had regular help so you were able to spend at least half of each summer camping with friends, playing in a softball league, and retaking a couple of college courses.

9. International Communication Assignment

There are many opportunities to work abroad, whether in internships, through a contracting firm, or through direct hiring. Using a Web site for a professional organization in your major field or a Web site such as www.internabroad.com, find an overseas internship or employment opportunity that interests you. Research and write a report on the cultural practices of the country in which the internship or employment is located. Your instructor will indicate whether your report will be oral or written. Information can be acquired from sources such as the following:

- Faculty and students who have visited the country
- Internet sites on other nations and on international communication
- Friends and colleagues familiar with the country you have chosen
- Books and articles on international communication and working overseas

Chapter | **11** | Style in Technical Writing

>>> Chapter Outline

This chapter, as well as the Handbook (Appendix A), focuses on the final stage of the writing process—revising. As you may already have discovered, revision sometimes gets short shrift during the rush to finish documents on time. That is a big mistake. Your writing must be clear, concise, and correct if you expect the reader to pay attention to your message. Toward that end, this chapter offers a few basic guidelines on style. The Handbook contains alphabetized entries on grammar, mechanics, and usage.

After defining style and its importance, this chapter gives suggestions for achieving five main stylistic goals:

- Writing clear sentences
- Being concise
- Being accurate in wording
- Using the active voice
- Using nonsexist language

>>> Overview of Style

This section (1) provides an overview of *style* as it applies to technical writing and (2) defines one particularly important aspect of style—called *tone*—that relates to every guideline in this chapter.

Definition of Style

Just as all writers have distinct personalities, they also display distinct features in their writing. Writing style can be defined as follows:

> **Style:** **The features of one's writing that show its individuality, separating it from the writing of others and shaping it to fit the needs of particular situations. Style results from the conscious and subconscious decisions each writer makes in matters like word choice, word order, sentence length, and active and passive voice. These decisions are different from the "right and wrong" matters of grammar and mechanics (see the Handbook). Instead, they are composed of choices writers make in deciding how to transmit ideas.**

Style is usually a series of personal decisions you make when you write. As noted in Chapter 2, however, much writing is being done these days by teams of writers. Collaborative writing requires individual writers to combine their efforts to produce a consensus style, usually a compromise of stylistic preferences of the individuals involved. Thus, personal style becomes absorbed into a jointly produced product. Similarly, many companies tend to develop a company style in documents like reports and proposals.

Importance of Tone

Tone is a major component of style and thus deserves special mention here. Through tone, you express an attitude in your writing—for example, neutral objectivity on the one hand, or unbridled enthusiasm on the other. The attitude evident in your tone exerts great influence over the reader. Indeed, it can determine whether your document achieves its objectives. Much like the broader term *style, tone* refers to the way you say something rather than what you say.

The following adjectives show a few examples of the types of tone or attitude that can be reflected in your writing. Here they are correlated with specific examples of documents:

1. **Casual tone:** E-mail to three colleagues working with you on a project.
2. **Objective tone:** Formal report to a client in which you present data comparing cost information for replacing the company's computer infrastructure.
3. **Persuasive tone:** Formal proposal to a client in hopes of winning a contract for goods or services.
4. **Enthusiastic tone:** Recommendation letter to a university to accept one of your employees in a master's program.
5. **Serious tone:** Memorandum to employees about the need to reduce the workforce and close an office.
6. **Authoritative tone:** Memo to an employee in which you reprimand him or her for violations of a policy about documenting absences.
7. **Friendly tone:** Letter to long-term clients inviting them to an open house at your new plant location.

Although there are almost as many variations in tone as there are occasions to write documents, one guideline always applies: Be as positive as you can possibly be, considering the context. Negative writing has little place in technical communication. In particular, a condescending or sarcastic tone should be avoided at all costs. It is the kind of writing you will regret. When you stress the positive, you stand the best chance of accomplishing your purpose and gaining the reaction you want from the reader.

Despite the need to make style conform to team or company guidelines, each individual remains the final arbiter of her or his own style in technical writing. Most of us will be our own stylists, even in firms in which in-house editors help clean up writing errors. This chapter helps such writers deal with everyday decisions of sentence arrangement, word choice, and the like. However, although style is a personal statement, you should not presume that anything goes. Certain fundamentals are part of all good technical style in the professional world. Let us take a look at these basics.

>>> Writing Clear Sentences

Each writer has his or her own approach to sentence style; yet, everyone has the same tools with which to work: words, phrases, and clauses. This section defines some basic terminology in sentence structure, and then it provides simple stylistic guidelines for writing clear sentences.

Sentence Terms

The most important sentence parts are the subject and verb. The *subject* names the person doing the action or the thing being discussed (e.g., *He* completed the study/The *figure* shows that); the *verb* conveys action or state of being (e.g., She *visited* the site/He *was* the manager).

Whether they are subjects, verbs, or other parts of speech, words are used in two main units: phrases and clauses. A *phrase* lacks a subject or verb or both and it thus must always relate to or modify another part of the sentence (She relaxed *after finishing her presentation./As project manager,* he had to write the report). A *clause,* however, has both a subject and a verb. Either it stands by itself as a *main clause (He talked to the team)* or it relies on another part of the sentence for its meaning and is thus a *dependent clause (After she left the site,* she went home).

Beyond these basic terms for sentence parts, you also should know the four main types of sentences:

- A *simple sentence* contains one main clause *(He completed his work).*

- A *compound sentence* contains two or more main clauses connected by conjunctions *(He completed his work, but she stayed at the office to begin another job).*

- A *complex sentence* includes one main clause and at least one dependent clause *(After he finished the project, he headed for home).*

- A *compound–complex sentence* contains at least two main clauses and at least one dependent clause *(After they studied the maps, they left the fault line, but they were unable to travel much farther that night).*

Guidelines for Sentence Style

Knowing the basic terms of sentence structure makes it easier to apply stylistic guidelines. Following are a few fundamental guidelines that form the underpinnings for good technical writing. As you review and edit your own writing or that of others, put these principles into practice.

>> Guideline 1: Place the Main Point Near the Beginning

One way to satisfy this criterion for good style is to avoid excessive use of the passive voice (see "Using the Active Voice" on pp. 231–232); another way is to avoid lengthy phrases or clauses at the beginnings of sentences. Remember that the reader usually wants the most important information first.

Original: "After reviewing the growth of the Cleveland office, it was decided by the corporate staff that an additional lab should be constructed at the Cleveland location."

Revision: "The corporate staff decided to build a new lab in Cleveland after reviewing the growth of the office there."

>> Guideline 2: Focus on One Main Clause in Each Sentence

When you string together too many clauses with *and* or *but,* you dilute the meaning of your text. However, an occasional compound or compound–complex sentence is acceptable, just for variety.

Original: "The hiring committee planned to interview Jim Steinway today, but bad weather delayed his plane departure, and the committee had to reschedule the interview for tomorrow."

Revision: "The hiring committee had to change Jim Steinway's interview from today to tomorrow because bad weather delayed his flight."

>> Guideline 3: Vary Sentence Length, but Seek an Average Length of 15–20 Words

Of course, do not inhibit your writing process by counting words while you write. Instead, analyze one of your previous reports to see how you fare. If your sentences are too long, make an effort to shorten them, such as by making two sentences out of one compound sentence connected by *and* or *but*.

You should also vary the length of sentences. Such variety keeps your reader's attention engaged. Make an effort to place important points in short but emphatic sentences. Reserve longer sentences for supporting main points.

Original: "Our field trip for the project required that we conduct research on Cumberland Island, a national wilderness area off the Georgia Coast, where we observed a number of species that we had not seen on previous field trips. Armadillos were common in the campgrounds, along with raccoons that were so aggressive that they would come out toward the campfire for a handout while we were still eating. We saw the wild horses that are fairly common on the island and were introduced there by explorers centuries ago, as well as a few bobcats that were introduced fairly recently in hopes of checking the expanding population of armadillos."

Revision: "Our field trip required that we complete research on Cumberland Island, a wilderness area off the Georgia Coast. There we observed many species we had not seen on previous field trips. Both armadillos and raccoons were common in the campgrounds. Whereas the armadillos were docile, the raccoons were quite aggressive. They approached the campfire for a handout while we were still eating. We also encountered Cumberland's famous wild horses, introduced centuries ago by explorers. Another interesting sighting was a pair of bobcats. They were brought to the island recently to check the expanding armadillo population."

>>> Being Concise

Some experts believe that careful attention to conciseness could shorten technical documents by 10% to 15%. As a result, reports and proposals would take less time to read and cost less to produce. This section on conciseness offers several techniques for reducing verbiage without changing meaning.

>> Guideline 1: Put Actions in Verbs

Concise writing depends more on verbs than it does on nouns. Sentences that contain abstract nouns that hide actions can be shortened by putting the action in strong verbs instead. By converting abstract nouns to action verbs, you can eliminate wordiness, as the following sentences illustrate:

Wordy: "The *acquisition* of the property was accomplished through long and hard negotiations."

Concise: "The property was *acquired* through long and hard negotiations."

Wordy: "*Confirmation* of the contract occurred yesterday."

Concise: "The contract was *confirmed* yesterday."

Wordy: "*Exploration* of the region had to be effected before the end of the year."

Concise: "The region had to be *explored* before the end of the year."

As the examples show, abstract nouns often end with *-tion* or *-ment* and are often followed by the preposition *of*. These words are not always "bad" words; they cause problems only when they replace action verbs from which they are derived. The following examples show some noun phrases along with the preferred verb substitutes:

assessment of	assess
classification of	classify
computation of	compute
delegation of	delegate
development of	develop
disbursement of	disburse
documentation of	document
elimination of	eliminate
establishment of	establish
negotiation of	negotiate
observation of	observe
requirement of	require
verification of	verify

>> Guideline 2: Shorten Wordy Phrases

Many wordy phrases have become common in business and technical writing. Weighty expressions add unnecessary words and rob prose of clarity. Following are some of the culprits, along with their concise substitutes:

afford an opportunity to	permit
along the lines of	like
an additional	another
at a later date	later
at this point in time	now
by means of	by

come to an end	end
due to the fact that	because
during the course of	during
for the purpose of	for
give consideration to	consider
in advance of	before
in the amount of	of
in the event that	if
in the final analysis	finally
in the proximity of	near
prior to	before
subsequent to	after
with regard to	about

>> Guideline 3: Replace Long Words with Short Ones

In grade school, most students are taught to experiment with long words. Although this effort helps build vocabularies, it also can lead to a lifelong tendency to use long words when short ones will do. Of course, sometimes you want to use longer words just for variety—for example, using an occasional *approximately* for the preferred *about*. As a rule, however, the following long words (in the left column) should routinely be replaced by the short words (in the right column):

advantageous	helpful
alleviate	lessen, lighten
approximately	about
cognizant	aware
commence	start, begin
demonstrate	show
discontinue	end, stop
endeavor	try
finalize	end, complete
implement	carry out
initiate	start, begin
inquire	ask
modification	change
prioritize	rank, rate
procure	buy
terminate	end, fire

transport	move
undertake	try, attempt
utilize	use

>> Guideline 4: Leave Out Clichés

Clichés are worn-out expressions that add words to your writing. Although they once were fresh phrases, they became clichés when they no longer conveyed their original meaning. You can make writing more concise by replacing clichés with a good adjective or two. Following are some clichés to avoid:

ballpark figure

efficient and effective

last but not least

needless to say

reinvent the wheel

skyrocketing costs

>> Guideline 5: Make Writing More Direct by Reading It Aloud

Much wordiness results from talking around a topic. Sometimes called *circumlocution,* this stylistic flaw arises from a tendency to write indirectly. It can be avoided by reading passages aloud. Hearing the sound of the words makes problems of wordiness quite apparent. It helps condense all kinds of inflated language, including the wordy expressions mentioned earlier. Remember, however, that direct writing must also retain a tactful and diplomatic tone when it conveys negative or sensitive information.

Indirect: "At the close of the last phase of the project, a bill for your services should be expedited to our central office for payment."

Direct: "After the project ends, please send your bill immediately to our central office."

Indirect: "It is possible that the well-water samples collected during our investigation of the well on the site of the subdivision could possibly contain some chemicals in concentrations higher than is allowable according to the state laws now in effect."

Direct: "Our samples from the subdivision's well might contain chemical concentrations beyond those permitted by the state."

>> Guideline 6: Avoid *There Are, It Is,* and Similar Constructions

There are and *it is* should not be substituted for concrete subjects and action verbs, which are preferable in good writing. Such constructions delay the delivery of information about who or what is doing something. They tend to make your writing lifeless and abstract. Avoid them by creating (1) main subjects that are concrete nouns and (2) main verbs that are action words. Note that the following revised passages give readers a clear idea of who is doing what in the subject and verb positions.

Original: "It is clear to the hiring committee that writing skills are an important criterion for every technical position."

Revision: "The hiring committee believes that writing skills are an important criterion for every technical position."

Original: "There were 15 people who attended the meeting at the client's office in Charlotte."

Revision: "Fifteen people attended the meeting at the client's office in Charlotte."

>> Guideline 7: Cut Out Extra Words

This guideline covers all wordiness errors not mentioned earlier. You must keep a vigilant eye for any extra words or redundant phrasing. Sometimes the problem comes in the form of needless connecting words, like *to be* or *that*. Other times it appears as redundant points—that is, those that have been made earlier in a sentence, paragraph, or section and do not need repeating.

Delete extra words when their use (1) does not add a necessary transition between ideas or (2) does not provide new information to the reader. (One important exception is the intentional repetition of main points for emphasis, as in repeating important conclusions in different parts of a report.)

>>> Being Accurate in Wording

Good technical writing also demands accuracy in phrasing. Technical professionals place their reputations and financial futures on the line with every document that goes out the door. That fact shows the importance of taking your time on editing that deals with the accuracy of phrasing. Accuracy often demands more words, not fewer. The main rule is:

Never sacrifice clarity for conciseness.

Careful writing helps to limit liability that your organization may incur. Your goal is very simple: Make sure words convey the meaning you intend—no more, no less. Some basic guidelines to follow include:

>> Guideline 1: Distinguish Facts from Opinions

In practice, this guideline means you must identify opinions and judgments as such by using phrases like *we recommend, we believe, we suggest,* or *in our opinion.*

Example: "In our opinion, spread footings would be an acceptable foundation for the building you plan at the site."

If you want to avoid repetitious use of such phrases, group your opinions into listings or report sections. Thus, a single lead-in can show the reader that opinions, not facts, are forthcoming.

Example: "On the basis of our site visit and our experience at similar sites, we believe that (1) _____, (2) _____, and (3) _____."

>> Guideline 2: Include Obvious Qualifying Statements When Needed

This guideline does not mean you must be overly defensive in every part of the report; it means that you must be wary of possible misinterpretations.

Example: "Our summary of soil conditions is based only on information obtained during a brief visit to the site. We did not drill any soil borings."

>> Guideline 3: Use Absolute Words Carefully

Avoid words that convey an absolute meaning or that convey a stronger meaning than you intend. One notable example is *minimize,* which means to reduce to the lowest possible level or amount. If a report claims that a piece of equipment will *minimize* breakdowns on the assembly line, the passage could be interpreted as an absolute commitment. The reader could consider any breakdown at all to be a violation of the report's implications. If instead the writer had used the verb *limit* or *reduce,* the wording would have been more accurate and less open to misunderstanding.

>>> Using the Active Voice

Striving to use the active voice can greatly improve your technical writing style. This section defines the active and passive voices and then gives examples of each. It also lists some practical guidelines for using both voices.

What Do Active and Passive Mean?

Active-voice sentences emphasize the person (or thing) performing the action—that is, somebody (or something) does something ("Matt completed the field study yesterday"). Passive-voice sentences emphasize the recipient of the action itself—that is, something is being done to something by somebody ("The field study was completed [by Matt] yesterday"). Following are some other examples of the same thoughts being expressed in first the active and then the passive voice:

■ **Examples:** Active-Voice Sentences:

1. "We *reviewed* aerial photographs in our initial assessment of possible fault activity at the site."

2. "The study *revealed* that three underground storage tanks had leaked unleaded gasoline into the soil."

3. "We *recommend* that you use a minimum concrete thickness of 6 in. for residential subdivision streets."

■ **Examples:** Passive-Voice Sentences:

1. "Aerial photographs *were reviewed* [by us] in our initial assessment of possible fault activity at the site."

2. "The fact that three underground storage tanks had been leaking unleaded gasoline into the soil *was revealed* in the study."

3. "*It is recommended* that you use a minimum concrete thickness of 6 in. for residential subdivision streets."

Just reading through these examples gives the sense that passive constructions are wordier than active ones. Also, passive voices tend to leave out the person or thing doing the action. Although occasionally this impersonal approach is appropriate, the reader can become frustrated by writing that fails to say who or what is doing something.

When Should Active and Passive Voices Be Used?

Both the active and passive voices have a place in your writing. Knowing when to use each is the key. Following are a few guidelines that will help:

- *Use the active voice when you want to:*

  1. Emphasize who is responsible for an action ("*We recommend* that you consider....")

  2. Stress the name of a company, whether yours or the reader's ("*PineBluff Contracting expressed* interest in receiving bids to perform work at....")

  3. Rewrite a top-heavy sentence so that the person or thing doing the action is up front ("*Figure 1 shows* the approximate locations of....")

  4. Pare down the verbiage in your writing, because the active voice is usually a shorter construction.

- *Use the passive voice when you want to:*

  1. Emphasize the object of the action or the action itself rather than the person performing the action ("*Samples will be sent* directly from the site to our laboratory in Sacramento").

  2. Avoid the kind of egocentric tone that results from repetitious use of *I, we,* and the name of your company ("*The project will be directed* by two programmers from our Boston office").

  3. Break the monotony of writing that relies too heavily on active-voice sentences.

Although the passive voice has its place, it is far too common in business and technical writing. This stylistic error results from the common misperception that passive writing is more objective. In fact, excessive use of the passive voice only makes writing more tedious to read. In modern business and technical writing, strive to use the active voice.

>>> Using Nonsexist Language

Language usually follows changes in culture rather than anticipating such changes. An example is today's shift away from sexist language in business and technical writing—indeed, in all writing and speaking. The change reflects the increasing number of women entering previously male-dominated professions, such as engineering, management, medicine, and law. It also reflects the fact that many men have taken previously female-dominated positions as nurses and flight attendants.

This section on style defines sexist and nonsexist language. Then it suggests ways to avoid using gender-offensive language in your writing.

Sexism and Language

Sexist language is the use of wording, especially masculine pronouns like *he* or *him,* to represent positions or individuals who could be either men or women. For many years, it was perfectly appropriate to use *he, his, him*, or other masculine words in sentences such as the following:

- "The operations specialist should check page 5 of his manual before flipping the switch."
- "Every physician was asked to renew his membership in the medical association before next month."
- "Each new student at the military academy was asked to leave most of his personal possessions in the front hallway of the administration building."

The masculine pronoun was understood to represent any person—male or female. Such usage came under attack for several reasons:

1. As previously mentioned, the entry of many more women into male-dominated professions has called attention to the inappropriate generic use of masculine pronouns.

2. Many people believe the use of masculine pronouns in a context that could include both genders constrains women from achieving equal status in the professions and in the culture—that is, the use of masculine pronouns encourages sexism in society as a whole.

Either point supplies a good enough reason to avoid sexist language. Many women in positions of responsibility may read your on-the-job writing. If you fail to rid your writing of sexism, you risk drawing attention toward sexist language and away from your ideas. Common sense argues for following some basic style techniques to avoid this problem.

Techniques for Nonsexist Language

This section offers techniques for shifting from sexist to nonsexist language. When shifting to nonsexist language, many writers have problems with subject–verb agreement. (The *engineer* recorded *their* data.) The strategies that follow help you avoid this problem. Not all these strategies will suit your taste in writing style; use the ones that work for you.

>> Technique 1: Avoid Personal Pronouns Altogether

One easy way to avoid sexist language is to delete or replace unnecessary pronouns:

Example:

Sexist Language: "During *his* first day on the job, any new employee in the toxic-waste laboratory must report to the company doctor for *his* employment physical."

| *Nonsexist Language:* | "During *the* first day on the job, each new employee in the toxic-waste laboratory must report to the company doctor for *a* physical." |

>> Technique 2: Use Plural Pronouns Instead of Singular

In most contexts you can shift from singular to plural pronouns without altering meaning. The plural usage avoids the problem of using masculine pronouns.

Example:

| *Sexist Language:* | "*Each* geologist should submit *his* time sheet by noon on the Thursday before checks are issued." |
| *Nonsexist Language:* | "*All* geologists should submit *their* time sheets on the Thursday before checks are issued." |

Interestingly, you may encounter sexist language that uses generic female pronouns inappropriately. For example, "Each nurse should make every effort to complete *her* rounds each hour." As in the preceding case, a shift to plural pronouns is appropriate: "Nurses should make every effort to complete *their* rounds each hour."

>> Technique 3: Alternate Masculine and Feminine Pronouns

Writers who prefer singular pronouns can avoid sexist use by alternating *he* and *him* with *she* and *her.* When using this technique, writers should avoid the unsettling practice of switching pronoun use within too brief a passage, such as a paragraph or page. Instead, writers may switch every few pages, or every section or chapter.

Although this technique is not yet in common use, its appeal is growing. It gives writers the linguistic flexibility to continue to use masculine and feminine pronouns in a generic fashion. However, one problem is that the alternating use of masculine and feminine pronouns tends to draw attention to itself. Also, the writer must work to balance the use of masculine and feminine pronouns, in a sense to give equal treatment.

>> Technique 4: Use Forms Like *He or She, Hers or His,* and *Him or Her*

This solution requires the writer to include pronouns for both genders.

Example:

| *Sexist Language:* | "The president made it clear that each branch manager will be responsible for the balance sheet of *his* respective office." |
| *Nonsexist Language:* | "The president made it clear that each branch manager will be responsible for the balance sheet of *his or her* respective office." |

This stylistic correction of sexist language may bother some readers. They believe that the doublet structure of *her or his,* is wordy and awkward. Many readers are bothered even more by the slash formations of *he/she, his/her,* and *her/him.* Avoid using these.

>> Technique 5: Shift to Second-Person Pronouns

Consider shifting to the use of *you* and *your,* words without any sexual bias. This technique is effective only with documents in which it is appropriate to use an instructions-related command tone associated with the use of *you*.

Example:

Sexist Language:	"After selecting *her* insurance option in the benefit plan, each new nurse should submit *her* paperwork to the Human Resources Department."
Nonsexist Language:	"Submit *your* paperwork to the Human Resources Department after selecting *your* insurance option in the benefit plan."

>> Technique 6: Be Especially Careful of Titles and Letter Salutations

Today, most women in business and industry are comfortable being addressed as *Ms.* If you know that the recipient prefers *Miss* or *Mrs.,* use that in your salutation. If a person's gender is not obvious from the name, call the person's employer and ask how the person prefers to be addressed. (When calling, also check on the correct spelling of the person's name and the person's current job title.) Receptionists and secretaries expect to receive such inquiries.

When you do not know who will read your letter, never use *Dear Sir* or *Gentlemen* as a generic greeting. Such a mistake may offend women reading the letter and may even cost you some business. *Dear Sir or Madam* is also inappropriate. It shows you do not know your audience, and it includes the archaic form *Madam.* Instead, call the organization for the name of a particular person to whom you can direct your letter. If you must write to a group of people, replace the generic greeting with an *Attention* line that denotes the name of the group.

Examples:

Sexist Language:	Dear Miss Finnegan: [to a single woman for whom you can determine no title preference]
Nonsexist Language:	Dear Ms. Finnegan:
Sexist Language:	Dear Sir: *or* Gentlemen:
Nonsexist Language:	Attention: Admissions Committee

No doubt the coming years will bring additional suggestions for solving the problem of sexist language. Whatever the culture finally settles on, it is clear that good technical writing style no longer tolerates the use of such language.

>>> Plain English and Simplified English

When you are writing technical or business documents, you may be asked to use one of two important styles of workplace writing: Plain English or Simplified English. Both of these styles include specific recommendations about sentence structure and word choice, but they are designed for particular audiences and purposes.

Plain English

Plain English is a specific style recommended for the U.S. government documents and for documents, such as proposals and reports that are submitted to federal agencies. Although people had been discussing clearer government documents for years, the Plain Language movement gained strong support during the mid-1990s. In 1995, a group of people began creating standards for Plain English in government writing. This group became the Plain Language Action and Information Network (PLAIN).

Plain English guidelines include many of the elements of clear technical communication: audience awareness, good document design, effective use of headings, and clear organization. However, Plain English is most clearly defined by its style recommendations, which include the following:

- Use active voice
- Put actions in strong verbs
- Use *you* to speak directly to the reader
- Use short sentences (no longer than about 20 words)
- Use concrete words
- Use simple and compound sentences with a subject–verb structure
- Make sure that modifiers are clear
- Use parallel structure for parallel ideas
- Avoid wordiness

The Plain Language Web site at http://www.plainlanguage.gov includes a complete discussion of Plain English with examples and links to other resources.

Simplified English

Simplified English includes many of the same recommendations as Plain English, and it is sometimes confused with Plain English. However, it serves a different purpose and is designed for a different audience. Simplified English, sometimes called *Controlled English* or *Internationalized English,* is designed for the global economy. It is designed for an audience for whom English is a second language, to be easily translatable from English into other languages. A leading organization for the development of Simplified English is the European Association of Aerospace Manufacturers (AECMA), which created the original standard in the 1980s.

Simplified English is designed to be clear and unambiguous, so it recommends specific sentence structures and limited vocabulary. Simplified English includes the following:

- Use only approved words
- Use one word for each meaning (avoid synonyms)
- Use only one meaning for each word (e.g., *close* is used only as a verb)
- Use active voice

- Use strong verbs
- Use articles (*a, an, the*) or demonstrative adjectives (*this, that, these, those*) for clarity
- Avoid strings of more than three nouns
- Use short sentences (less than 20 words)

More information about Simplified English standards is available at http://www.asd-ste100.org, and an overview of Simplified English is available at http://www.userlab.com/SE.html, which also includes a sample list of approved words at http://www.userlab.com/Downloads/SE.pdf. Because the standards were developed for the aerospace industry, the word lists are specialized for that industry. Other industries are developing their own word lists. A more general word list can be downloaded from the Publications/Documents section of Intecom's Web site at http://www.intecom.org.

>>> Chapter Summary

Style is an important part of technical writing. During the editing process, writers make the kinds of changes that place their personal stamp on a document. Style can also be shaped (1) by a team, in that writing done collaboratively can acquire features of its diverse contributors or (2) by an organization, in that an organization may require writers to adopt a particular writing style. Yet the decision-making process of individual writers remains the most important influence on the style of technical documents.

This chapter offers five basic suggestions for achieving good technical writing style. First, sentences should be clear, with main ideas at the beginning and with one main clause in most sentences. Although sentences should average only 15 to 20 words, you should vary sentence patterns in every document. Second, technical writing should be concise. You can achieve this goal by reading prose aloud as you rewrite and edit. Third, wording should be accurate. Fourth, the active voice should be dominant, although the passive voice also has a place in good technical writing. And fifth, the language of technical documents should be free of sexual bias.

Although clarity and conciseness are important to all workplace writing, you may be asked to follow a specific style sheet. Many organizations create their own in-house style guides, and some styles, such as Plain English and Simplified English are used for industries or to clarify cross-cultural communication.

>>> Learning Portfolio

Collaboration at Work Describing *Style*

General Instructions

Each Collaboration at Work exercise applies strategies for working in teams to chapter topics. The exercise assumes you (1) have been divided into teams of about three to six students, (2) will use team time inside or outside of class to complete the case, and (3) will produce an oral or written response. For guidelines about writing in teams, refer to Chapter 2.

Background for Assignment

As this chapter points out, the term *style* refers to the way you choose to express an idea, as opposed to the content of the idea itself. The definition of *style* early in this chapter makes it clear that writers adopt particular styles for different contexts. For example, following are three passages that express the same idea in three different ways:

1. The results of the experiment strongly suggest to the team conducting the study that the hypothesis is valid.
2. After evaluating the results of the experiment, we concluded that the hypothesis is valid.
3. We believe the experiment worked.

Team Assignment

Describe how the previous three passages convey information differently to the reader. Can you describe the differences in style? When is one passage more appropriate to use than another?

Assignments

1. Conciseness—Abstract Words

Make the following sentences more concise by replacing abstract nouns with verbs. Other minor changes in wording may be necessary.

a. The inspectors indicated that observation of the site occurred on July 16, 2011.

b. After three hours of discussion, the branch managers agreed that establishment of a new mission statement should take place in the next fiscal year.

c. Assessment of the firm's progress will happen during the annual meeting of the Board of Directors.

d. Documentation of the results of the lab test appeared in the final report.

e. The financial statement showed that computation of the annual revenues had been done properly.

2. Conciseness—Wordy Phrases and Long Words

Condense the following sentences by replacing long phrases and words with shorter substitutes.

a. In the final analysis, we decided to place the new pumping station in the proximity of the old one.

b. Endeavoring to complete the study on time, Sheila transported the supplies immediately from the field location to the lab.

c. During the course of his career, he planned to utilize the experience he had gained in the ambulance business.

d. His work with the firm terminated due to the fact that he took a job with another competing firm.

e. Subsequent to the announcement he made, he held a news conference for approximately one hour of time.

3. Conciseness—Clichés and *There Are/It Is* Constructions

Rewrite the following sentences by eliminating clichés and the wordy constructions *there are* and *it is*.

a. There are many examples of skyrocketing equipment costs affecting the final budget for a project.

b. She explained to her staff that it was as plain as day that they would have to decrease their labor costs.

c. The prospective client asked for a ballpark figure of the project costs.

d. Last but not least, there was the issue of quality control that he wanted to emphasize in his speech.

e. It is a fact that our boss ended the meeting about a loss of profits by noting that we are all in the same boat.

4. Sentence Clarity

Improve the clarity of the following sentences by changing sentence structures or by splitting long sentences into several shorter ones.

a. Therefore, to collect a sample from above the water table, and thus to follow the directions provided by the client, the initial boring was abandoned and the drill rig was repositioned about two feet away and a new boring was drilled.

b. Percolation test #1 was performed approximately 40 feet east of the existing pump house and percolation test #2 was performed near the base of the slope approximately 65 feet west of the pump house, and then the results were submitted to the builder.

c. All of the earth materials encountered in our exploration can be used for trench backfill above manhole and pipe bedding, provided they are free of organic material, debris, and other deleterious materials, and they are screened to remove particles greater than six inches in diameter.

d. The properties consist of approximately 5,000 acres, including those parcels of Heron Ranch owned by American Axis Insurance Company, the unsold Jones Ranch parcels, the village commercial area, the mobile home subdivisions, two condominium complexes, a contractor's storage area, an RV storage area, a sales office, a gatehouse, open space parcels, and the undeveloped areas for future Buildings 1666, 1503, 1990, and 1910.

e. Having already requested permits for the construction of the bathhouse, medical center, maintenance building, boat dock, swimming pool, community building, and an addition to the community building, we still need to apply for the storeroom permit.

5. Active and Passive Voice Verbs

Make changes in active and passive voice verbs, where appropriate. Refer to the guidelines in the chapter. Be able to supply a rationale for any change you make.

a. It was recommended by the personnel committee that you consider changing the requirements for promotion.

b. No formal report about assets was reported by the corporation before it announced the merger.

c. It has been noted by the Department of Environmental Services that the laundry business was storing toxic chemicals in an unsafe location.

d. The violation of ethical guidelines was reported by the commissioner to the president of the association.

e. Dirt brought to the site should be evaluated by the engineer on-site before it is placed in the foundation.

6. Sexist Language

Revise the following sentences to eliminate sexist language.

a. Although each manager was responsible for his own budget, some managers obviously had better accounting skills than others.

b. Each flight attendant is required to meet special work standards as long as she is employed by an international airline.

c. Typically, a new engineer at Blue Sky receives his first promotion after about a year.

d. Every worker wonders whether he is saving enough for retirement.

e. If a pilot senses danger, she should abort the takeoff.

7. Advanced Exercise—Conciseness

The following sentences contain more words than necessary. Rewrite each passage more concisely but without changing the meaning. If appropriate, make two sentences out of one.

a. The disbursement of the funds from the estate will occur on the day that the proceedings concerning the estate are finalized in court.

b. At a later date we plan to begin the process of prioritizing our responsibilities on the project so that we will have a clear idea of which activities deserve the most attention from the project personnel.

c. Needless to say, we do not plan to add our participation to the project if we conclude that the skyrocketing costs of the project will prohibit our earning what could be considered to be a fair profit from the venture.

d. For us to supply the additional supplies that the client wishes to procure from us, the client will have to initiate a change order that permits additional funds to be transferred into the project account.

e. Prior to the implementation of the state law with regard to the use of asbestos as a building material, it was common practice to utilize this naturally occurring mineral in all kinds of facilities, some of which became health hazards subsequently.

8. Advanced Exercise—General Style Rules

Revise the following sentences by applying all the guidelines mentioned in this chapter. When you change passive verbs to active, it may be necessary to make some assumptions about the agent of the action, because the sentences are taken out of context.

a. Based on our review of the available records, conversations with the various agencies involved, including the Fire Department and the Police Department, and a thorough survey of the site where the spill occurred, it was determined that the site contained chemicals that were hazardous to human health.

b. It is recommended by us that your mainframe computer system be replaced immediately by a newer model.

c. The figures on the firm's profit margins in July and August, along with sales commissions for the last six months of the previous year and the top 10 salespersons in the firm, are included in the Appendix.

d. It was suggested by the team that the company needs to invest in modern equipment.

e. It is the opinion of this writer that the company's health plan is adequate.

f. Shortly after the last change in leadership, and during the time that the board of directors was expressing strong views about the direction that the company was taking, it became clear to me and other members of the senior staff that the company was in trouble.

g. Each manager should complete and submit his monthly report by the second Tuesday of every month.

9. Editing Paper of Classmate

For this assignment, exchange papers with a member of your class. Use either the draft of a current assignment or a paper that was completed earlier in the term. Edit your classmate's work in accordance with this chapter's guidelines on style, and then explain your changes to the writer.

10. Editing Sample Memo

Using the guidelines in this chapter, edit the following memorandum. The assignment can be completed individually or in teams as a team-editing project.

DATE: January 12, 2011
TO: All Employees of Denver Branch
FROM: Leonard Schwartz
             Branch Manager
SUBJECT: New Loss-Prevention System

As you may have recently heard, lately we received news from the corporate headquarters of the company that it would be in the best interest of the entire company to pay more attention to matters of preventing accidents and any other safety-related measures that affect the workplace, including both office and field activities related to all types of jobs that we complete. Every single employee in each office at every branch needs to be ever mindful in this regard so that he is most efficient and effective in the daily performance of his everyday tasks that relate to his job responsibilities such that safety is always of paramount concern.

With this goal of safety ever present in our minds, I believe the bottom line of the emphasis on safety could be considered to be the training that each of us receives in his first, initial weeks on the job as well as the training provided on a regular basis throughout each year of our employment with Whitman Development, so that we are always aware of how to operate in a safe manner. The training vehicle gives the company the mechanism to provide each of you with the means to become aware of the elements of safety that relate to the specific needs and requirements of your own particular job. Therefore, at this point in time I have come to the conclusion in the process of contemplating the relevance of the new corporate emphasis on safety to our particular branch that we need, as a branch, to give much greater scrutiny and analysis to the way we can prevent accidents and emphasize the concern of safety at every stage of our operation for every employee. Toward this end, I have asked the training coordinator, Kendra Jones, to assemble a written training program that will involve every single employee and that can be implemented beginning no later than June of this year. When the plan has been written and approved at the various levels within the office, I will conduct a meeting with every department in order to emphasize the major and minor components of this upcoming safety program.

It is my great pleasure to announce to all of you that effective in the next month (February) I will give a monthly safety award of $100 to the individual branch employee at any level of the branch who comes up with the best, most useful suggestion related to safety in any part of the branch activities. Today I will take the action of placing a suggestion box on the wall of the lunchroom so that all of you will have easy access to a way to get your suggestions for safety into the pipeline and to be considered. As an attachment to the memo you are now reading from me, I have provided you with a copy of the form that you are to use in making any suggestions that are then to be placed in the suggestion box. On the last business day of each month, the box will be emptied of the completed forms for that month, and before the end of the following week a winner will be selected by me for the previous month's suggestion program and an announcement will be placed by me to that effect on the bulletin board in the company workroom.

　　If you have any questions in regard to the corporate safety program as it affects our branch or about the suggestion program that is being implemented here at the Denver office at Whitman, please do not hesitate to make your comments known either in memorandum form or by way of telephonic response to this memorandum.

11. Ethics Assignment

Assume that you are an electrical engineer at a civil engineering firm. Evelyn, the technical writer in your office, has let engineers know that she is available to help with articles and presentations. You have asked for her help in preparing an article for publication in a professional journal. As you hand her the article, you are quick to add that you have long-standing problems organizing information and editing well. Two days later the draft appears in your mailbox looking like your first graded paper in English 101 in college. Evelyn has even provided a suggested outline for reorganizing the entire piece. On reading her comments and reviewing the outline, you find that you agree with almost all of her suggestions. You follow her suggestions and proceed to meet with her several times and show her three more drafts, including the final that she edits and proofs.

　　Feeling that she has done more on your article than she would normally do as part of her job responsibilities, Evelyn diplomatically asks how you plan to acknowledge her work on the final published article. How do you respond to her? Do you list her on the title page as coauthor, do you mention her in a footnote as an editor, or do you adopt some other approach? Explain the rationale you give Evelyn after telling her your decision. What are the main ethical considerations in making the decision?

12. International Communication Assignment

One major problem with international communication occurs when product instructions are written (or translated) by individuals who do not have enough familiarity with the language being used. The problem can be solved by *localization*, or choosing writers or translators who are, in fact, native speakers and writers. For this assignment, locate a set of instructions written in English with stylistic errors that would not have been made by a native speaker/writer. Point out these errors and suggest appropriate revisions.

Appendix A

>>> Handbook

This handbook includes entries on the basics of writing. It contains three main types of information:

1. **Grammar:** the rules by which we edit sentence elements. Examples include rules for the placement of punctuation, the agreement of subjects and verbs, and the placement of modifiers.
2. **Mechanics:** the rules by which we make final proofreading changes. Examples include the rules for abbreviations and the use of numbers. A list of commonly misspelled words is also included.
3. **Usage:** information on the correct use of particular words, especially pairs of words that are often confused. Examples include problem words like *affect / effect, complement / compliment,* and *who / whom.*

This handbook is presented in alphabetized fashion for easy reference during the editing process. Grammar and mechanics entries are in all uppercase; usage entries are in lowercase. Several exercises follow the entries.

A/An

A and *an* are different forms of the same article. *A* occurs before words that start with consonants or consonant sounds. EXAMPLES:

- a three-pronged plug
- a once-in-a-lifetime job (*once* begins with the consonant sound of *w*)
- a historic moment (many speakers and some writers mistakenly use *an* before *historic*)

An occurs before words that begin with vowels or vowel sounds. EXAMPLES:

- an eager new employee
- an hour before closing

A lot/Alot

The correct form is the two-word phrase *a lot.* Although acceptable in informal discourse, *a lot* usually should be replaced by more formal diction in technical writing. EXAMPLE: "They retrieved many [*not a lot of*] soil samples from the construction site."

Abbreviations

Technical writing uses many abbreviations. Without this shorthand form, you end up writing much longer reports and proposals without any additional content. Use the following seven basic rules in your use of abbreviations, paying special attention to the first three:

Rule 1: Do Not Use Abbreviations When Confusion May Result

When you want to use a term just once or twice and you are not certain your readers will understand an abbreviation, write out the term rather than abbreviating it. EXAMPLE: "They were required to remove creosote from the site, according to the directive from the Environmental Protection Agency." Even though *EPA* is the accepted abbreviation for this government agency, you should write out the name in full if you are using the term only once to an audience that may not understand it.

Rule 2: Use Parentheses for Clarity

When you use a term more than twice and are not certain that your readers will understand it, write out the term the first time it is used and place the abbreviation in parentheses, and then use the abbreviation in the rest of the document. In long reports or proposals, however, you may need to repeat the full term in key places. EXAMPLE: "According to the directive from the Environmental Protection Agency (EPA), they were required to remove the creosote from the construction site. Furthermore, the directive indicated that the builders could expect to be visited by EPA inspectors every other week."

Rule 3: Include a Glossary When There Are Many Abbreviations

When your document contains many abbreviations that may not be understood by all readers, include a well-marked glossary at the beginning or end of the document. A glossary simply collects all the terms and abbreviations and places them in one location for easy reference.

Rule 4: Use Abbreviations for Units of Measure

Most technical documents use abbreviations for units of measure. Do not include a period unless the abbreviation could be confused with a word. EXAMPLES: mi, ft, oz, gal., in., and lb. Note that units-of-measurement abbreviations have the same form for both singular and plural amounts. EXAMPLES: ½ in., 1 in., 5 in.

Rule 5: Avoid Spacing and Periods

Avoid internal spacing and internal periods in most abbreviations that contain all capital letters. EXAMPLES: ASTM, EPA, ASEE. Exceptions include professional titles and degrees, such as P.E., B.S., and B.A.

Rule 6: Be Careful with Company Names

Abbreviate a company or other organizational name only when you are sure that officials from the organization consider the abbreviation appropriate. IBM (for the company) and UCLA (for the university) are examples of commonly accepted organizational abbreviations. When in doubt, follow rule 2—write the name in full the first time it is used, followed by the abbreviation in parentheses.

Rule 7: Common Abbreviations

The following common abbreviations are appropriate for most writing in your technical or business career. They are placed into three main categories of measurements, locations, and titles.

Measurements.   Use these abbreviations only when you place numbers before the measurement.

ac	alternating current	gpm	gallons per minute
amp	ampere	hp	horsepower
bbl	barrel	hr	hour
Btu	British thermal unit	Hz	hertz
bu	bushel	in.	inch
C	Celsius	j	joule
cal	calorie	K	Kelvin
cc	cubic centimeter	ke	kinetic energy
circ	circumference	kg	kilogram
cm	centimeter	km	kilometer
cos	cosine	kw	kilowatt
cot	cotangent	kwh	kilowatt-hour
cps	cycles per second	l	liter
cu ft	cubic feet	lb	pound
db	decibel	lin	linear
dc	direct current	lm	lumen
dm	decimeter	log.	logarithm
doz *or* dz	dozen	m	meter
F	Fahrenheit	min	minute
f	farad	mm	millimeter
fbm	foot board measure	oz	ounce
fig.	figure	ppm	parts per million
fl oz	fluid ounce	psf	pounds per square foot
FM	frequency modulation	psi	pounds per square inch
fp	foot pound	pt	pint
ft	foot (feet)	qt	quart
g	gram	rev	revolution
gal.	gallon	rpm	revolutions per minute

sec	second		va	volt-ampere
sq	square		w	watt
sq ft	square foot (feet)		wk	week
T	ton		wl	wavelength
tan.	tangent		yd	yard
v	volt		yr	year

Locations.   Use these common abbreviations for addresses (e.g., on envelopes, letters and resumes), but write out the words in full in other contexts.

AL	Alabama		MT	Montana
AK	Alaska		NE	Nebraska
AS	American Samoa		NV	Nevada
AZ	Arizona		NH	New Hampshire
AR	Arkansas		NJ	New Jersey
CA	California		NM	New Mexico
CZ	Canal Zone		NY	New York
CO	Colorado		NC	North Carolina
CT	Connecticut		ND	North Dakota
DE	Delaware		OH	Ohio
DC	District of Columbia		OK	Oklahoma
FL	Florida		OR	Oregon
GA	Georgia		PA	Pennsylvania
GU	Guam		PR	Puerto Rico
HI	Hawaii		RI	Rhode Island
ID	Idaho		SC	South Carolina
IL	Illinois		SD	South Dakota
IN	Indiana		TN	Tennessee
IA	Iowa		TX	Texas
KS	Kansas		UT	Utah
KY	Kentucky		VT	Vermont
LA	Louisiana		VI	Virgin Islands
ME	Maine		VA	Virginia
MD	Maryland		WA	Washington
MA	Massachusetts		WV	West Virginia
MI	Michigan		WI	Wisconsin
MN	Minnesota		WY	Wyoming
MS	Mississippi		Alta.	Alberta
MO	Missouri		B.C.	British Columbia

Man.	Manitoba		Ont.	Ontario
N.B.	New Brunswick		P.E.I.	Prince Edward Island
Nfld.	Newfoundland		P.Q.	Quebec
N.W.T.	Northwest Territories		Sask.	Saskatchewan
N.S.	Nova Scotia		Yuk.	Yukon

Titles. Some of the following abbreviations go before the name (e.g., Dr., Ms., Messrs.), whereas others go after the name (e.g., college degrees, Jr., Sr.).

Atty.	Attorney		M.A.	Master of Arts
B.A.	Bachelor of Arts		M.S.	Master of Science
B.S.	Bachelor of Science		M.D.	Doctor of Medicine
D.D.	Doctor of Divinity		Messrs.	Plural of Mr.
Dr.	Doctor (used mainly with medical and dental degrees but also with other doctorates)		Mr.	Mister
			Mrs.	Used to designate married, widowed, or divorced women
Drs.	Plural of Dr.		Ms.	Used increasingly for all women, especially when one is uncertain about a woman's marital status
D.V.M.	Doctor of Veterinary Medicine			
Hon.	Honorable			
Jr.	Junior		Ph.D.	Doctor of Philosophy
LL.D.	Doctor of Laws		Sr.	Senior

Accept/Except

Accept and *except* have different meanings and often are different parts of speech. *Accept* is a verb that means "to receive." *Except* is a preposition or verb and means "to make an exception or special case of." EXAMPLES:

- I *accepted* the service award from my office manager.
- Everyone *except* Jonah attended the marine science lecture.
- The company president *excepted* me from the meeting because I had an important sales call to make the same day.

Advice/Advise/Inform

Advice is a noun that means "suggestions or recommendation." *Advise* is a verb that means "to suggest or recommend." Do not use the verb *advise* as a substitute for *inform*, which means simply "to provide information." EXAMPLES:

- The consultant gave us *advice* on starting a new retirement plan for our employees.
- She *advised* us that a 401(k) plan would be useful for all our employees.
- She *informed* [not *advised*] her clients that they would receive her final report by March 15.

Affect/Effect

Affect and *effect* generate untold grief among many writers. The key to using them correctly is remembering two simple sentences: (1) *affect* with an *a* is a verb meaning "to influence" and (2) *effect* with an *e* is a noun meaning "result." There are some exceptions, however, such as when *effect* can be a verb that means "to bring about," as in, "He effected considerable change when he became a manager." EXAMPLES:

- His progressive leadership greatly *affected* the company's future.
- One *effect* of securing the large government contract was the hiring of several more accountants.
- The president's belief in the future of microcomputers *effected* change in the company's approach to office management. (For a less-wordy alternative, substitute *changed* for *effected change in*.)

Agree to/Agree with

In correct usage, *agree to* means that you have *consented to* an arrangement, an offer, a proposal, and so on. *Agree with* is less constraining and only suggests that you are *in harmony with* a certain statement, idea, person, and the like. EXAMPLES:

- Representatives from BoomCo *agreed to* alter the contract to reflect the new scope of work.
- We *agree with* you that more study may be needed before the nuclear power plant is built.

All Right/Alright

All right is an acceptable spelling; *alright* is not. *All right* is an adjective that means "acceptable," an exclamation that means "outstanding," or a phrase that means "correct." EXAMPLES:

- Sharon suggested that the advertising copy was *all right* for now but that she would want changes next month.
- Upon seeing his article in print, Zach exclaimed, *"All right!"*
- The five classmates were *all right* in their response to the trick questions on the quiz.

All Together/Altogether

All together is used when items or people are being considered in a group or are working in concert. *Altogether* is a synonym for "utterly" or "completely." EXAMPLES:

- The three firms were *all together* in their support of the agency's plan.
- There were *altogether* too many pedestrians walking near the dangerous intersection.

Allusion/Illusion/Delusion/Elusion

These similar sounding words have distinct meanings. Following is a summary of the differences:

1. **Allusion:** a noun meaning "reference," as in you are making an allusion to your vacation in a speech. The related verb is *allude*.
2. **Illusion:** a noun meaning "misunderstanding or false perception." It can be physical (as in seeing a mirage) or mental (as in having the false impression that your hair is not thinning when it is).
3. **Delusion:** a noun meaning "a belief based on self-deception." Unlike *illusion*, the word conveys a much stronger sense that someone is out of touch with reality, as in having "delusions of grandeur." The related verb is *delude*.
4. **Elusion:** a noun meaning "the act of escaping or avoiding." The more common form is the verb, *elude*, meaning "to escape or avoid."

Examples:
- His report included an *allusion* to the upcoming visit by the government agency in charge of accreditation.
- She harbored an *illusion* that she was certain to receive the promotion. In fact, her supervisor preferred another department member with more experience.
- He had *delusions* that he soon would become company president, even though he started just last week in the mailroom.
- The main point of the report *eluded* him because there was no executive summary.

Already/All Ready

All ready is a phrase that means "everyone is prepared," whereas *already* is an adverb that means something is finished or completed. EXAMPLES:

- They were *all ready* for the presentation to the client.
- George had *already* arrived at the office before the rest of his proposal team members had even left their homes.

Alternately/Alternatively

Because many readers are aware of the distinction between these two words, any misuse can cause embarrassment or even misunderstanding. Follow these guidelines for correct use.

Alternately. As a derivative of *alternate, alternately* is best reserved for events or actions that occur "in turns." EXAMPLE: While digging the trench, he used a backhoe and a hand shovel *alternately* throughout the day.

Alternatively. A derivative of *alternative, alternatively* should be used in contexts where two or more choices are being considered. EXAMPLE: We suggest that you use deep foundations at the site. *Alternatively*, you could consider spread footings that were carefully installed.

Amount/Number

Amount is used in reference to items that *cannot* be counted, whereas *number* is used to indicate items that *can* be counted. EXAMPLES:

- In the last year, we have greatly increased the *amount* of computer paper ordered for the Boston office.
- The last year has seen a huge increase in the *number* [not *amount*] of boxes of computer paper ordered for the Boston office.

And/Or

This awkward expression probably has its origins in legal writing. It means that there are three separate options to be considered: the item before *and/or*, the item after *and/or*, or both items.

Avoid *and/or* because readers may find it confusing, visually awkward, or both. Instead, replace it with the structure used in the previous sentence; that is, write "A, B, or both," *not* "A and/or B." EXAMPLE:

The management trainee was permitted to select two seminars from the areas of computer hardware, communication skills, or both [not *computer hardware and/or communication skills*].

Anticipate/Expect

Anticipate and *expect* are not synonyms. In fact, their meanings are distinctly different. *Anticipate* is used when you mean to suggest or state that steps have been taken beforehand to prepare for a situation. *Expect* only means you consider something likely to occur. EXAMPLES:

- *Anticipating* that the contract will be successfully negotiated, Jones Engineering is hiring three new hydrologists.
- We *expect* [not *anticipate*] that you will encounter semicohesive and cohesive soils in your excavations at the Park Avenue site.

Apt/Liable/Likely

Maintain the distinctions in these three similar words.

1. *Apt* is an adjective that means "appropriate," "suitable," or "has an aptitude for."
2. *Liable* is an adjective that means "legally obligated" or "subject to."
3. *Likely* is either an adjective that means "probable" or "promising" or an adverb that means "probably." As an adverb, it should be preceded by a qualifier such as *quite*.

Examples:

- The successful advertising campaign showed that she could select an *apt* phrase for selling products.
- Jonathan is *apt* at running good meetings. He always hands out an agenda and always ends on time.

- The contract makes clear who is *liable* for any on-site damage.
- Completing the warehouse without an inspection will make the contractor *liable* to lawsuits from the owner.
- A *likely* result of the investigation will be a change in the law. [*likely* as an adjective]
- The investigation will quite *likely* result in a change in the law. [*likely* as an adverb]

Assure/Ensure/Insure

Assure is a verb that can mean "to promise." It is used in reference to people, as in, "We want to *assure* you that our crews will strive to complete the project on time." In fact, *assure* and its derivatives (like *assurance*) should be used with care in technical contexts, because these words can be viewed as a guarantee.

The synonyms *ensure* and *insure* are verbs meaning "to make certain." Like *assure*, they imply a level of certainty that is not always appropriate in engineering or the sciences. When their use is deemed appropriate, the preferred word is *ensure;* reserve *insure* for sentences in which the context is insurance. EXAMPLES:

- Be *assured* that our representatives will be on-site to answer questions that the subcontractor may have.
- To *ensure* that the project stays within schedule, we are building in 10 extra days for bad weather. (An alternative: "So that the project stays within schedule, we are building in 10 extra days for bad weather.")

Augment/Supplement

Augment is a verb that means to increase in size, weight, number, or importance. *Supplement* is either (1) a verb that means "to add to" something to make it complete or to make up for a deficiency or (2) a noun that means "the thing that has been added." EXAMPLES:

- The power company supervisor decided to *augment* the line crews in five counties.
- He *supplemented* the audit report by adding the three accounting statements.
- The three accounting *supplements* helped support the conclusions of the audit report.

Awhile/A While

Though similar in meaning, this pair is used differently. *Awhile* means "for a short time." Because "for" is already a part of its definition, it cannot be preceded by the preposition "for." The noun *while*, however, can be preceded by the two words "for a," giving it essentially the same meaning as *awhile*. EXAMPLES:

- Kirk waited *awhile* before trying to restart the generator.
- Kirk *waited for a while* before trying to restart the generator.

Balance/Remainder/Rest

Balance should be used as a synonym for *remainder* only in the context of financial affairs. *Remainder* and *rest* are synonyms to be used in other nonfinancial contexts. EXAMPLES:

- The account had a *balance* of $500, which was enough to avoid a service charge.
- The *remainder* [or *rest,* but not *balance*] of the day will be spent on training in oral presentations for proposals.
- During the *rest* [not *balance*] of the session, we learned about the new office equipment.

Because/Since

Maintain the distinction between these two words. *Because* establishes a cause–effect relationship, whereas *since* is associated with time. EXAMPLES:

- *Because* he left at 3 P.M., he was able to avoid rush hour.
- *Since* last week, her manufacturing team completed 3,000 units.

Between/Among

The distinction between these two words has become somewhat blurred. However, many readers still prefer to see *between* used with reference to only two items, reserving *among* for three or more items. EXAMPLES:

- The agreement was just *between* my supervisor and me. No one else in the group knew about it.
- The proposal was circulated *among* all members of the writing team.
- *Among* Sallie, Todd, and Fran, there was little agreement about the long-term benefits of the project.

Bi-/Semi-/Biannual/Biennial

The prefixes *bi* and *semi* can cause confusion. Generally, *bi* means "every two years, months, weeks, etc.," whereas *semi* means "twice a year, month, week, etc." Yet many readers get confused by the difference, especially when they are confronted with a notable exception, such as *biannual* (which means twice a year) and *biennial* (which means every two years).

Your goal, as always, is clarity for the reader. Therefore, it is best to write out meanings in clear prose, rather than relying on prefixes that may not be understood. EXAMPLES:

- We get paid twice a month [preferable to *semimonthly* or *biweekly*].
- The part-time editor submits articles every other month [preferable to *bimonthly*].
- We hold a company social gathering twice a year [preferable to *biannually* or *semiannually*].
- The auditor inspects our safety files every two years [preferable to *biennially*].

Capital/Capitol

Capital is a noun whose main meanings are (1) a city or town that is a government center, (2) wealth or resources, or (3) net worth of a business or the investment that has been made in the business by owners. *Capital* can also be an adjective meaning (1) "excellent," (2) "primary," or (3) "related to the death penalty." Finally, *capital* can be a noun or an adjective referring to uppercase letters.

Capitol is a noun or an adjective that refers to a building where a legislature meets. With a capital letter, it refers exclusively to the building in Washington, D.C., where the U.S. Congress meets. EXAMPLES:

- The *capital* of Pickens County is Jasper, Georgia.
- Our family *capital* was reduced by the tornado and hurricane.
- She had invested significant *capital* in the carpet factory.
- Their proposal contained some *capital* ideas that would open new opportunities for our firm.
- In some countries, armed robbery is a *capital* offense.
- The students visited the *capitol* building in Atlanta. Next year they will visit the *Capitol* in Washington, D.C., where they will meet several members of Congress.

Capitalization

As a rule, you should capitalize *specific* names of people, places, and things—sometimes called *proper nouns*. For example, capitalize specific streets, towns, trademarks, geologic eras, planets, groups of stars, days of the week, months of the year, names of organizations, holidays, and colleges. However, remember that excessive capitalization—as in titles of positions in a company—is inappropriate in technical writing and can appear somewhat pompous.

The following rules cover some frequent uses of capitals:

1. Major words in titles of books and articles. Capitalize prepositions and articles only when they appear as the first word in titles. EXAMPLES:
   - *For Whom the Bell Tolls*
   - *In Search of Excellence*
   - *The Power of Positive Thinking*
2. Names of places and geographic locations. EXAMPLES:
   - Washington Monument
   - Cleveland Stadium
   - Dallas, Texas
   - Cobb County
3. Names of aircraft and ships. EXAMPLES:
   - *Air Force One*
   - *SS Arizona*
   - *Nina, Pinta,* and *Santa Maria*

4. Names of specific departments and offices within an organization. EXAMPLES:

- Humanities Department
- Personnel Department
- International Division

5. Political, corporate, and other titles that come before names. EXAMPLES:

- Chancellor Hairston
- Councilwoman Jones
- Professor Gainesberg
- Congressman Buffett

Note, however, that general practice does not call for capitalizing most titles when they are used by themselves or when they follow a person's name. EXAMPLES:

- Jane Cannon, a professor in the Business Department.
- Zachary Alan Mar, president of Alan Security.
- Chris Presley, secretary of the Oil Rig Division.

Center On/Revolve Around

The key to using these phrases correctly is to think about their literal meaning. For example, you center *on* (not around) a goal, just as you would center on a target with a gun or bow and arrow. Likewise, your hobbies revolve *around* your early interest in water sports, just as the planets revolve around the sun in our solar system. EXAMPLES:

- All her selling points in the proposal *centered on* the need for greater productivity in the factory.
- At the latest annual meeting, some stockholders argued that most of the company's recent projects *revolved around* the CEO's interest in attracting attention from the media.

Cite/Site/Sight

1. *Cite* is a verb meaning "to quote as an example, authority, or proof." It can also mean "to commend" or "to bring before a court of law" (as in receiving a traffic ticket).

2. *Site* usually is a noun that means "a particular location." It can also be a verb that means "to place at a location," as with a new school being sited by the town square, but this usage is not preferred. Instead use a more conventional verb, such as *built*.

3. *Sight* is a noun meaning "the act of seeing" or "something that is seen," or it can be a verb meaning "to see or observe."

Examples:

- We *cited* a famous geologist in our report on the earthquake.
- Rene was *cited* during the ceremony for her exemplary service to the city of Roswell.
- The officer will *cite* the party-goers for disturbing the peace.

- Although five possible dorm *sites* were considered last year, the college administrators decided to build [preferred over *site*] the dorm at a different location.
- The *sight* of the flock of whooping cranes excited the visitors.
- Yesterday we *sighted* five whooping cranes at the marsh.

Complement/Compliment

Both words can be nouns and verbs, and both have adjective forms (*complementary, complimentary*).

Complement.   *Complement* is used as a noun to mean "that which has made something whole or complete," as a verb to mean "to make whole, to make complete," or as an adjective. You may find it easier to remember the word by recalling its mathematical definition: Two *complementary* angles must always equal 90 degrees. EXAMPLES:

- As a noun: The *complement* of five technicians brought our crew strength up to 100%.
- As a verb: The firm in Canada served to *complement* ours in that together we won a joint contract.
- As an adjective: Seeing that project manager and her secretary work so well together made clear their *complementary* relationship in getting the office work done.

Compliment.   *Compliment* is used as a noun to mean "an act of praise, flattery, or admiration," as a verb to mean "to praise, to flatter," or as an adjective to mean "related to praise or flattery, or without charge." EXAMPLES:

- As a noun: He appreciated the verbal *compliments*, but he also hoped they would result in a substantial raise.
- As a verb: Howard *complimented* the crew for finishing the job on time and within budget.
- As an adjective: We were fortunate to receive several *complimentary* copies of the new software from the publisher.

Compose/Comprise

These are both acceptable words, with an inverse relationship to each other. *Compose* means "to make up or be included in," whereas *comprise* means "to include or consist of." The easiest way to remember this relationship is to memorize one sentence: "The parts compose the whole, but the whole comprises the parts." One more point to remember: The common phrase *is comprised of* is a substandard, unacceptable replacement for *comprise* or *is composed of*. Careful writers do not use it. EXAMPLES:

- Seven quite discrete layers *compose* the soils that were uncovered at the site.
- The borings revealed a stratigraphy that *comprises* [not *is comprised of*] seven quite discrete layers.

Consul/Council/Counsel

Consul, council, and *counsel* can be distinguished by meaning and, in part, by their use within a sentence.

> *Consul:* A noun meaning an official of a country who is sent to represent that country's interests in a foreign land.
>
> *Council:* A noun meaning an official group or committee.
>
> *Counsel:* A noun meaning an adviser or advice given, or a verb meaning to produce advice.

Examples:

- (Consul) The Brazilian *consul* met with consular officials from three other countries.
- (Council) The Human Resources *Council* of our company recommended a new retirement plan to the company president.
- (Counsel—as noun) After the tragedy, they received legal *counsel* from their family attorney and spiritual *counsel* from their minister.
- (Counsel—as verb) As a communications specialist, Roberta helps *counsel* employees who are involved in various types of disputes.

Continuous/Continual

The technical accuracy of some reports may depend on your understanding of the difference between *continuous* and *continual*. *Continuous* and *continuously* should be used in reference to uninterrupted, unceasing activities. However, *continual* and *continually* should be used with activities that are intermittent, or repeated at intervals. If you think your reader may not understand the difference, you should either (1) use synonyms that are clearer (such as *uninterrupted* for *continuous*, and *intermittent* for *continual*) or (2) define each word at the point you first use it in the document. EXAMPLES:

- We *continually* checked the water pressure for three hours before the equipment arrived, while also using the time to set up the next day's tests.
- Because it rained *continuously* from 10:00 A.M. until noon, we were unable to move our equipment onto the utility easement.

Criterion/Criteria

Coming from the Latin language, *criterion* and *criteria* are the singular and plural forms of a word that means "rationale or reasons for selecting a person, place, thing, or idea." A common error is to use *criteria* as both a singular and plural form, but such misuse disregards a distinction recognized by many readers. Maintain the distinction in your writing. EXAMPLES:

- Among all the qualifications we established for the new position, the most important *criterion* for success is good communication skills.
- She had to satisfy many *criteria* before being accepted into the honorary society of her profession.

Data/Datum

Coming as it does from the Latin, the word *data* is the plural form of *datum*. Although many writers now accept *data* as singular or plural, traditionalists in the technical and scientific community still consider *data* exclusively a plural form. Therefore, you should maintain the plural usage. EXAMPLES:

- These *data* show that there is a strong case for building the dam at the other location.
- This particular *datum* shows that we need to reconsider recommendations put forth in the original report.

If you consider the traditional singular form of *datum* to be awkward, use substitutes such as, "This item in the data shows..." or "One of the data shows that..." Singular subjects like *one* or *item* allow you to keep your original meaning without using the word *datum*.

Definite/Definitive

Although similar in meaning, these words have slightly different contexts. *Definite* refers to that which is precise, explicit, or final. *Definitive* has the more restrictive meaning of "authoritative" or "final." EXAMPLES:

- It is now *definite* that he will be assigned to the London office for six months.
- He received the *definitive* study on the effect of the oil spill on the marine ecology.

Discrete/Discreet/Discretion

The adjective *discrete* suggests something that is separate or made up of many separate parts. The adjective *discreet* is associated with actions that require caution, modesty, or reserve. The noun *discretion* refers to the quality of being "discreet," or the freedom a person has to act on her or his own. EXAMPLES:

- The orientation program at Antelope Equipment includes a writing seminar, which is a *discrete* training unit offered for one full day.
- The orientation program at Antelope Equipment includes five *discrete* units.
- As a counselor in Antelope Equipment's Human Resources Office, Sharon was *discreet* in her handling of personal information about employees.
- Every employee in the Human Resources Office was instructed to show *discretion* in handling personal information about employees.
- By starting a flextime program, Antelope Equipment will give employees a good deal of *discretion* in selecting the time to start and end their workday.

Disinterested/Uninterested

In contemporary business use, *disinterested* and *uninterested* have quite different meanings. Because errors can cause confusion for the reader, make sure not to use the words as synonyms.

Disinterested means "without prejudice or bias," whereas *uninterested* means "showing no interest." EXAMPLES:

■ The agency sought a *disinterested* observer who had no stake in the outcome of the trial.

■ They spent several days talking to officials from Iceland, but they still remain *uninterested* in performing work in that country.

Due to/Because of

Besides irritating those who expect proper English, mixing these two phrases can also cause confusion. *Due to* is an adjective phrase meaning "attributable to" and almost always follows a "to be" verb (such as "is," "was," or "were"). It should not be used in place of prepositional phrases, such as "because of," "owing to," or "as a result of." EXAMPLES:

■ The cracked walls were *due to* the lack of proper foundation fill being used during construction.

■ We won the contract *because of* [not *due to*] our thorough understanding of the client's needs.

Each Other/One Another

Each other occurs in contexts that include only two persons, whereas *one another* occurs in contexts that include three or more persons. EXAMPLES:

■ Shana and Katie worked closely with *each other* during the project.

■ All six members of the team conversed with *one another* regularly through e-mail.

e.g./i.e.

The abbreviation *e.g.* means "for example," whereas *i.e.* means "that is." These two Latin abbreviations are often confused, a fact that should give you pause before using them. Many writers prefer to write them out, rather than risk confusion on the part of the reader. EXAMPLES:

■ During the trip, he visited 12 cities where Max Entertainment is considering opening offices—e.g., [or, *for example*] Kansas City, New Orleans, and Seattle.

■ A spot along the Zayante Fault was the earthquake's epicenter—*i.e.*, [or *that is*] the focal point for seismic activity.

English as a Second Language (ESL)

Technical writing challenges native English speakers and nonnative speakers alike. The purpose of this section is to present a basic description of three grammatical forms: articles, verbs, and prepositions. These forms may require more intense consideration from international students when they complete technical writing assignments. Each issue is described using the ease-of-operation section from a memo about a fax machine. The passage, descriptions, and charts work together to show how these grammar issues function collectively to create meaning.

Ease of Operation—Article Usage

The AIM 500 is so easy to operate that **a** novice can learn to transmit **a** document to another location in about two minutes. Here's **the** basic procedure:

1. Press **the** button marked TEL on **the** face of **the** fax machine. You then hear **a** dial tone.
2. Press **the** telephone number of **the** person receiving **the** fax on **the** number pad on **the** face of **the** machine.
3. Lay **the** document face down on **the** tray at **the** back of **the** machine.

At this point, just wait for **the** document to be transmitted—about 18 seconds per page to transmit. **The** fax machine will even signal **the** user with **a** beep and **a** message on its LCD display when **the** document has been transmitted. Other more advanced operations are equally simple to use and require little training. Provided with **the** machine are two different charts that illustrate **the** machine's main functions.

The size of **the** AIM 500 makes it easy to set up almost anywhere in **an** office. **The** dimensions are 13 inches in width, 15 inches in length, and 9.5 inches in height. **The** narrow width, in particular, allows **the** machine to fit on most desks, file cabinets, or shelves.

Articles.   Articles are one of the most difficult forms of English grammar for nonnative English speakers, mainly because some language systems do not use them. Thus speakers of particular languages may have to work hard to incorporate the English article system into their language proficiency.

The English articles include *a, an,* and *the.*

- *A* and *an* express indefinite meaning when they refer to nouns or pronouns that are not specific. The writer believes the reader does not know the noun or pronoun.
- *The* expresses definite meaning when it refers to a specific noun or pronoun. The writer believes the reader knows the specific noun or pronoun.

ESL writers choose the correct article only when they (1) know the context or meaning, (2) determine whether they share information about the noun with the reader, and (3) consider the type of noun following the article.

The ease-of-operation passage includes 31 articles that represent two types—definite and indefinite. When a writer and a reader share knowledge of a noun, the definite article should be used. On 25 occasions the articles in the passage suggest the writer and reader share some knowledge of a count noun. *Count nouns* are nouns that can be counted (pen, cloud, memo). Examples of non-count nouns are sugar, air, and beef.

For example, the memo writer and the memo recipient share knowledge of the particular model fax machine—the AIM 500. Thus, *the* is definite when it refers to "the fax machine" in the memo. Notice, however, that *document* becomes definite only after the second time it is mentioned ("Lay the document face down…."). In the first reference to *document,* *a* document refers to a document about which the writer and reader share no knowledge. The memo writer cannot know which document the reader will fax. Only in the second reference do the writer and reader know the document to be the one the reader will fax.

The indefinite article *a* occurs five times, whereas *an* occurs once. Each occurrence signals a singular count noun. The reader and the writer share no knowledge of the nouns that follow the *a* or *an,* so an indefinite article is appropriate. *A* precedes nouns beginning with consonant sounds. *An* precedes nouns beginning with vowel sounds. Indefinite articles seldom precede non-count nouns unless a non-count functions as a modifier (a beef shortage).

Definite and indefinite articles are used more frequently than other articles; however, other articles do exist. The "generic" article refers to classes or groups of people, objects, and ideas. If the fax machine is thought of in a general sense, the meaning changes. For example, "the fax machine increased office productivity by 33%." *The* now has a generic meaning representing fax machines in general. The same generic meaning can apply to the plural noun, but such generic use requires no article: "Fax machines increased office productivity by 33%." *The* in this instance is a generic article.

Articles from "Ease of Operation" Excerpt

Article	Noun	Type	Comment
The	ATM 500	definite	first mention—shared knowledge
a	novice	indefinite	first mention—no shared knowledge
a	document	indefinite	first mention—no shared knowledge
the	basic procedure	definite	
the	button	definite	
the	face	definite	
the	fax machine	definite	first mention without proper name, with reader/writer shared knowledge
a	dial tone	indefinite	first mention—no shared knowledge
the	telephone number	definite	
the	person	definite	
the	fax	definite	
the	number pad	definite	
the	face	definite	
the	machine	definite	
the	document	definite	
the	tray	definite	
the	back	definite	
the	machine	definite	
the	document	definite	second mention
the	fax machine	definite	
the	user	definite	
a	beep	indefinite	first mention—no shared knowledge
a	message	indefinite	first mention—no shared knowledge
the	document	definite	
the	machine	definite	
the	machine's main	definite	functions
The	size	definite	
the	AIM 500	definite	
an	office	indefinite	first mention—preceding vowel sound—no shared knowledge
The	dimensions	definite	
The	narrow width	definite	
the	machine	definite	definite

Ease of Operation—Verb Usage

The AIM 500 is so easy to operate that a novice **can learn** to transmit a document to another location in about two minutes. Here's the basic procedure:

1. **Press** the button marked TEL on the face of the fax machine. You then **hear** a dial tone.
2. **Press** the telephone number of the person receiving the fax on the number pad on the face of the machine.
3. **Lay** the document face down on the tray at the back of the machine.

At this point, just **wait** for the document to be transmitted—about 18 seconds per page to transmit. The fax machine **will** even **signal** the user with a beep and a message on its LCD display when the document **has been transmitted**. Other more advanced operations **are** equally simple to use and **require** little training. **Provided** with the machine **are** two different charts that **illustrate** the machine's main functions.

The size of the AIM 500 **makes** it easy to set up almost anywhere in an office. The dimensions **are** 13 inches in width, 15 inches in length, and 9.5 inches in height. The narrow width, in particular, **allows** the machine to fit on most desks, file cabinets, or shelves.

Verbs. Verbs express time in three ways—simple present, simple past, and future. *Wait, waited,* and *will wait* and *lay* ("to put"), *laid,* and *will lay* are examples of simple present, simple past, and future tense verbs. Verbs in the English language system appear as either regular or irregular forms.

Regular Verbs—Simple Tense Regular verbs follow a predictable pattern. The form of the simple present tense verbs (*walk*) changes to the simple past tense with the addition of—*ed* (*walked*) and changes to the simple future with the addition of a special auxiliary (helping) verb called a *modal* (*will walk*).

Present	Past	Future
learn	learned	will learn
wait	waited	will wait
press	pressed	will press
signal	signaled	will signal
require	required	will require
provide	provided	will provide
illustrate	illustrated	will illustrate
allow	allowed	will allow

Irregular Verbs—Simple Tense Irregular verbs do not follow a predictable pattern. Most importantly, the past tense is not created by adding –*ed.* The simple present tense of *lay* ("to put") changes completely in the simple past (*laid*).

Present	Past	Future
is	was	will be
are	were	will be
hear	heard	will hear
do	did	will do
get	got	will get
see	saw	will see
write	wrote	will write
speak	spoke	will speak

Unfortunately, the English verb system is more complicated than that. Verbs express more than time; they can also express *aspect*, or whether an action was completed. The perfect aspect indicates that an action was completed (perfected) and the progressive aspect indicates that an action is incomplete (in progress).

Regular Verbs—Aspect   In regular verbs, the perfect aspect is indicated with the addition of a form of the auxiliary (helping) word *to have* to the simple past tense form. In verb phrases that indicate aspect, tense is always found in the first verb in the verb phrase. For example, "I have walked" is present perfect, and "I had walked" is past perfect. The progressive aspect is indicated with the addition of a form of the auxiliary word *to be* and an —*ing* form of the main verb. The progressive aspect is always regular.

Present Perfect	Past Perfect	Future Perfect
have learned	had learned	will have learned
have waited	had waited	will have waited
have pressed	had pressed	will have pressed
have signaled	had signaled	will have signaled
have required	had required	will have required
have provided	had provided	will have provided
have illustrated	had illustrated	will have illustrated
have allowed	had allowed	will have allowed
Present Progressive	**Past Progressive**	**Future Progressive**
is learning	was learning	will be learning
is waiting	was waiting	will be waiting
is pressing	was pressing	will be pressing
is signaling	was signaling	will be signaling
is requiring	was requiring	will be requiring
is providing	was providing	will be providing
is illustrating	was illustrating	will be illustrating
is allowing	was allowing	will be allowing

Irregular Verbs—Aspect   The irregular forms of the perfect aspect can be confusing. The auxiliary verbs are the same as for the regular verb phrases, but the main verb can be inflected in a number of ways. Most dictionaries list this form of the verb after the present and past forms of the verb.

Present Perfect	Past Perfect	Future Perfect
have been	had been	will have been
have been	had been	will have been
have heard	had heard	will have heard
have done	had done	will have done
have gotten	had gotten	will have gotten
have seen	had seen	will have seen
have written	had written	will have written
have spoken	had spoken	will have spoken

Let's examine four specific verb forms in the "Ease of Operation" passage.

1. *Is* represents a being or linking verb in the passage. Being verbs suggest an aspect of an experience or being (existence); for example, "He is still here," and "The fax is broken." Linking verbs connect a subject to a complement (completer); for example, "The fax machine is inexpensive."

2. *Can learn* is the present tense of the verb *learn* preceded by a modal. Modals assist verbs to convey meaning. *Can* suggests ability or possibility. Other modals and their meanings appear next.

Will	Would	Could	Shall	Should	Might	Must
scientific fact possibility determination	hypothetical	hypothetical	formal will	expectation obligation	possibility	necessity

3. *Here's* shows a linking verb (*is*) connected to its complement (*here*). The sentence in its usual order—subject first followed by the verb—appears as, "The basic procedure is here." Article—adjective—noun—linking verb—complement.

Verbs from "Ease of Operation" Excerpt

Verb	Tense	Number	Other Details
is	present	singular	linking/being (is, was, been)
can learn	present	singular	*can* is a modal auxiliary implying "possibility"
Here's (is)	present	singular	linking/being
Press	present	singular	understood "you" as subject
hear	present	singular	action/transitive
Press	present	singular	understood "you" as subject
Lay	present	singular	irregular (lay, laid, laid) singular—understood "you" as subject
wait	present	singular	understood "you" as subject
will signal	future	singular	action to happen or condition to experience
has been transmitted	present perfect	singular	passive voice—action that began in the past and continues to the present

(continued)

Verb	Tense	Number	Other Details
are	present	plural	linking/being
require	present	plural	action/transitive
Provided are	present perfect	plural	passive voice—action that began in the past and continues to the present
illustrate	present	plural	action/transitive
makes	present	singular	action/transitive
are	present	plural	linking/being
allows	present	singular	action/transitive

4. *Press, Lay,* and *wait* (for) share at least four common traits: present tense, singular number, action to transitive, and understood subject of "you." Although "you" does not appear in the text, the procedure clearly instructs the person operating the fax machine—"you." Action or transitive verbs express movement, activity, and momentum, and may take objects. Objects answer the questions Who? What? To whom? Or, for whom? in relation to transitive verbs. For example, "Press the button," "Hear a dial tone," "Press the telephone number," "Lay the document face down." Press What? Hear what? Lay what?

Ease of Operation—Preposition Usage

The AIM 500 is so easy to operate that a novice can learn to transmit a document **to** another location **in** about two minutes. Here's the basic procedure:

1. Press the button marked TEL **on** the face **of** the fax machine. You then hear a dial tone.
2. Press the telephone number **of** the person receiving the fax **on** the number pad **on** the face **of** the machine.
3. Lay the document face down **on** the tray **at** the back **of** the machine.

At this point, just wait **for** the document to be transmitted—**about** 18 seconds **per** page to transmit. The fax machine will even signal the user **with** a beep and a message **on** its LCD display when the document has been transmitted. Other more advanced operations are equally simple to use and require little training. Provided **with** the machine are two different charts that illustrate the machine's main functions.

The size **of** the AIM 500 makes it easy to set up almost anywhere **in** an office. The dimensions are 13 inches **in** width, 15 inches **in** length, and 9.5 inches **in** height. The narrow width, **in** particular, allows the machine to fit **on** most desks, file cabinets, or shelves.

Prepositions.   Prepositions are words that become a part of a phrase composed of the preposition, a noun or pronoun, and any modifiers. Notice the relationships expressed within the prepositional phrases and the ways they affect meaning in the sentences. In the "Ease of Operation" passage, about half the prepositional phrases function as adverbs noting place or time; the other half function as adjectives.

Place or	Location	Time
at	on	before
in	above	after
below	around	since
beneath	out	during
over	underneath	
within	under	
outside	near	
into	inside	

One important exception is a preposition that connects to a verb to make a *prepositional verb*—*wait for*. Another interesting quality of prepositions is that sometimes more than one can be used to express similar meaning. In the "Ease of Operation" passage, for example, both *on* the tray and *at* the back indicate position. Another way to state the same information is *on* the tray *on* the back.

Prepositions from "Ease of Operation" Excerpt

Preposition	Noun Phrase	Comment
to	another location	direction toward
in	(about) two minutes	approximation of time
on	the face	position
of	the fax machine	originating at or from
of	the person	associated with
on	the number pad	position
on	the face	position
of	the machine	originating at
on	the tray	position
at	the back	position of
of	the machine	originating at
At	this point	on or near the time
for	the document	indication of object of desire
about	18 seconds	adverb = approximation
per	page	for every
with	a beep and a message	accompanying
on	its LCD display	position
with	the machine	accompanying
of	the AIM 500	originating at or from
in	an office	within the area
in	width	with reference to
in	length	with reference to
in	height	with reference to
in	particular	with reference to
on	most desks, file cabinets, or shelves	position

Farther/Further

Although similar in meaning, these two words are used differently. *Farther* refers to actual physical distance, whereas *further* refers to nonphysical distance or can mean "additional." EXAMPLES:

- The overhead projector was moved *farther* from the screen so that the print would be easier to see.
- *Farther* up the old lumber road, they found footprints of an unidentified mammal.
- As he read *further* along in the report, he began to understand the complexity of the project.
- She gave *further* instructions after they arrived at the site.

Fewer/Less

The adjective *fewer* is used before items that can be counted, whereas the adjective *less* is used before mass quantities. When errors occur, they usually result from *less* being used with countable items, as in this *incorrect* sentence: "We can complete the job with less men at the site." EXAMPLES:

- The newly certified industrial hygienist signed with us because the other firm in which he was interested offered *fewer* [not *less*] benefits.
- There was *less* sand in the sample taken from 15 ft than in the one taken from 10 ft.

Flammable/Inflammable/Nonflammable

Given the importance of these words in avoiding injury and death, make sure to use them correctly—especially in instructions. *Flammable* means "capable of burning quickly" and is acceptable usage. *Inflammable* has the same meaning, but it is not acceptable usage for this reason: Some readers confuse it with *nonflammable*. The word *nonflammable*, then, means "not capable of burning" and is accepted usage. EXAMPLES:

- They marked the package *flammable* because its contents could be easily ignited by a spark. (Note that *flammable* is preferred here over its synonym, *inflammable*.)
- The foreman felt comfortable placing the crates near the heating unit, because all the crates' contents were *nonflammable*.

Former/Latter

These two words direct the reader's attention to previous points or items. *Former* refers to that which came first, whereas *latter* refers to that which came last. Note that the words are used together when there are only two items or points—not with three or more. Also, you should know that some readers may prefer you avoid *former* and *latter* altogether, because the construction may force them to look back to previous sentences to understand your meaning. The second example gives an alternative.

- (with former/latter) The airline's machinists and flight attendants went on strike yesterday. The *former* left work in the morning, whereas the *latter* left work in the afternoon.
- (without former/latter) The airline's machinists and flight attendants went on strike yesterday. The machinists left work in the morning, whereas the flight attendants left work in the afternoon.

Fortuitous/Fortunate

The word *fortuitous* is an adjective that refers to an unexpected action, without regard to whether it is desirable. The word *fortunate* is an adjective that indicates an action that is clearly desired. The common usage error with this pair is the wrong assumption that *fortuitous* events must also be *fortunate*. EXAMPLES:

- Seeing Digital Essential's London manager at the conference was quite *fortuitous*, because I had not been told that he also was attending.
- It was indeed *fortunate* that I encountered the London manager, for it gave us the chance to talk about an upcoming project involving both our offices.

Generally/Typically/Usually

Words like *generally, typically,* and *usually* can be useful qualifiers in your reports. They indicate to the reader that what you have stated is often, but not always, the case. Make certain to place these adverb modifiers as close as possible to the words they modify. In the first example, it would be inaccurate to write *were typically sampled,* because the adverb modifies the entire verb phrase *were sampled.* EXAMPLES:

- Cohesionless soils *typically* were sampled by driving a 2-in.-diameter, split-barrel sampler. (Active-voice alternative: *Typically*, we sampled cohesionless soils by driving a 2-in.-diameter, split-barrel sampler.)
- For projects like the one you propose, the technician *usually* cleans the equipment before returning to the office.
- It is *generally* known that sites for dumping waste should be equipped with appropriate liners.

Good/Well

Although similar in meaning, *good* is used as an adjective and *well* is used as an adverb. A common usage error occurs when writers use the adjective when the adverb is required. EXAMPLES:

- It is *good* practice to submit three-year plans on time.
- He did *well* to complete the three-year plan on time, considering the many reports he had to finish that same week.

Imply/Infer

Remember that the person doing the speaking or writing implies, whereas the person hearing or reading the words infers. In other words, the word *imply* requires an active role; the word *infer* requires a passive role. When you *imply* a point, your words suggest rather than state a point. When you *infer* a point, you form a conclusion or deduce meaning from someone else's words or actions. EXAMPLES:

- The contracts officer *implied* that there would be stiff competition for that $20 million waste-treatment project.
- We *inferred* from her remarks that any firm hoping to secure the work must have completed similar projects recently.

Its/It's

Its and *it's* are often confused. You can avoid error by remembering that *it's* with the apostrophe is used *only* as a contraction for *it is* or *it has*. The other form—*its*—is a possessive pronoun. You can remember this by remembering that other possessive pronouns (mine, his) do not have apostrophes. EXAMPLES:

- Because of the rain, *it's* [or *it is*] going to be difficult to move the equipment to the site.
- *It's* [or *it has*] been a long time since we submitted the proposal.
- The company completed *its* part of the agreement on time.

Lay/Lie

Lay and *lie* are troublesome verbs, and you must know some basic grammar to use them correctly.

1. *Lay* means "to place." It is a transitive verb; thus it takes a direct object to which it conveys action. ("She laid down the printout before starting the meeting.") Its main forms are *lay* (present), *laid* (past), *laid* (past participle), and *laying* (present participle).

2. *Lie* means "to be in a reclining position." It is an intransitive verb; thus it does not take a direct object. ("In some countries, it is acceptable for workers to lie down for a midday nap.") Its main forms are *lie* (present), *lay* (past), *lain* (past participle), and *lying* (present participle).

If you want to use these words with confidence, remember the transitive/intransitive distinction and memorize the principal parts. EXAMPLES:

- (lay) I will *lay* the notebook on the lab desk before noon.
- (lay) I have *laid* the notebook there before.
- (lay) I was *laying* the notebook down when the phone rang.
- (lie) The watchdog *lies* motionless at the warehouse gate.

- (lie) The dog *lay* there yesterday too.
- (lie) The dog has *lain* there for three hours today and no doubt will be *lying* there when I return from lunch.

Lead/Led

Lead is either a noun that names the metallic element or a verb that means "to direct or show the way." *Led* is only a verb form, the past tense of the verb *lead*. EXAMPLES:

- The company bought rights to mine *lead* on the land.
- They chose a new president to *lead* the firm into the twenty-first century.
- They were *led* to believe that salary raises would be high this year.

Like/As

Like and *as* are different parts of speech and thus are used differently in sentences. *Like* is a preposition and therefore is followed by an object—not an entire clause. *As* is a conjunction and thus is followed by a group of words that includes a verb. *As if* and *as though* are related conjunctions. EXAMPLES:

- Gary looks *like* his father.
- Managers *like* John will be promoted quickly.
- If Teresa writes this report *as* she wrote the last one, our clients will be pleased.
- Our proposals are brief, *as* they should be.
- Our branch manager talks *as though* [or *as if*] the merger will take place soon.

Loose/Lose

Loose, which rhymes with "goose," is an adjective that means "unfastened, flexible, or unconfined." *Lose*, which rhymes with "ooze," is a verb that means "to misplace." EXAMPLES:

- The power failure was linked to a *loose* connection at the switchbox.
- Because of poor service, the photocopy machine company may *lose* its contract with Digital Essential's San Francisco office.

Modifiers: Dangling and Misplaced

This section includes guidelines for avoiding the most common modification errors—dangling modifiers and misplaced modifiers. First, however, we must define the term *modifier*. Words, phrases, and even dependent clauses can serve as modifiers. They serve to qualify, or add meaning to, other elements in the sentence. For our purposes here, the most important point is that modifiers must be connected clearly to what they modify.

Modification errors occur most often with verbal phrases. A phrase is a group of words that lacks either a subject or a predicate. The term *verbal* refers to (1) gerunds

(–*ing* form of verbs used as nouns, such as, "He likes skiing"), (2) participles (–*ing* form of verbs used as adjectives, such as, "Skiing down the hill, he lost a glove"), or (3) infinitives (the word *to* plus the verb root, such as, "To attend the opera was his favorite pastime"). Now let's look at the two main modification errors.

Dangling Modifiers.   When a verbal phrase "dangles," the sentence in which it is used contains no specific word for the phrase to modify. As a result, the meaning of the sentence can be confusing to the reader. For example, "In designing the foundation, several alternatives were discussed." It is not at all clear exactly who is doing the "designing." The phrase dangles because it does not modify a specific word. The modifier does not dangle in this version of the sentence: "In designing the foundation, we discussed several alternatives."

Misplaced Modifiers.   When a verbal phrase is misplaced, it may appear to refer to a word that it, in fact, does not modify. EXAMPLE: "Floating peacefully near the oil rig, we saw two humpback whales." Obviously, the whales are doing the floating, and the rig workers are doing the seeing here. Yet because the verbal phrase is placed at the beginning of the sentence, rather than at the end immediately after the word it modifies, the sentence presents some momentary confusion.

Misplaced modifiers can lead to confusion about the agent of action in technical tasks. EXAMPLE: "Before beginning to dig the observation trenches, we recommend that the contractors submit their proposed excavation program for our review." On quick reading, the reader is not certain about who will be "beginning to dig"—the contractors or the "we" in the sentence. The answer is the contractors. Thus a correct placement of the modifier should be, "We recommend the following: Before the contractors begin digging observation trenches, they should submit their proposed excavation for our review."

Solving Modifier Problems.   At best, dangling and misplaced modifiers produce a momentary misreading by the audience. At worst, they can lead to confusion that results in disgruntled readers, lost customers, or liability problems. To prevent modification problems, place all verbal phrases—indeed, all modifiers—as close as possible to the word they modify. If you spot a modification error while you are editing, correct it in one of two ways:

1. Leave the modifier as it is and rework the rest of the sentence. Thus you would change "Using an angle of friction of 20 degrees and a vertical weight of 300 tons, the sliding resistance would be…" to the following: "Using an angle of friction of 20 degrees and a vertical weight of 300 tons, we computed a sliding resistance of…."

2. Rephrase the modifier as a complete clause. Thus you would change the previous original sentence to, "If the angle of friction is 20 degrees and the vertical weight is 300 tons, the sliding resistance should be…."

In either case, your goal is to link the modifier clearly and smoothly with the word or phrase it modifies.

Number of/Total of

These two phrases can take singular or plural verbs, depending on the context. Following are two simple rules for correct usage:

1. If the phrase is preceded by *the*, it takes a singular verb because emphasis is placed on the group.
2. If the phrase is preceded by *a*, it takes a plural verb because emphasis is placed on the many individual items.

Examples:

- *The number of* projects going over budget *has* decreased dramatically.
- *The total of* 90 lawyers *believes* the courtroom guidelines should be changed.
- *A number* of projects *have* stayed within budget recently.
- *A total of* 90 lawyers *believe* the courtroom guidelines should be changed.

Numbers

Like rules for abbreviations, those for numbers vary from profession to profession and even from company to company. Most technical writing subscribes to the approach that numbers are best expressed in figures (45) rather than words (forty-five). Note that this style may differ from that used in other types of writing, such as this textbook. Unless the preferences of a particular reader suggest that you do otherwise, follow these common rules for use of numbers in writing your technical documents:

Rule 1: Follow the 10-or-Over Rule

In general, use figures for numbers of 10 or more, words for numbers under 10. EXAMPLES: Three technicians at the site/15 reports submitted last month/one rig contracted for the job.

Rule 2: Do Not Start Sentences with Figures

Begin sentences with the word form of numbers, not with figures. EXAMPLE: "Forty-five containers were shipped back to the lab."

Rule 3: Use Figures as Modifiers

Whether higher or lower than 10, numbers are usually expressed as figures when used as modifiers with units of measurement, time, and money, especially when these units are abbreviated. EXAMPLES: 4 in., 7 hr, 17 ft, $5 per hr. Exceptions can be made when the unit is not abbreviated. EXAMPLE: five years.

Rule 4: Use Figures in a Group of Mixed Numbers

Use only figures when the numbers grouped together in a passage (usually *one* sentence) are both higher and lower than 10. EXAMPLE: "For that project they assembled 15 samplers, 4 rigs, and 25 containers." In other words, this rule argues for consistency within a writing unit.

Rule 5: Use the Figure Form in Illustration Titles

Use the numeric form when labeling specific tables and figures in your reports. EXAMPLES: Figure 3, Table 14–B.

Rule 6: Be Careful with Fractions

Express fractions as words when they stand alone, but as figures when they are used as a modifier or are joined to whole numbers. EXAMPLE: "We have completed two-thirds of the project using the 2½;-in. pipe."

Rule 7: Use Figures and Words with Numbers in Succession

When two numbers appear in succession in the same unit, write the first as a word and the second as a figure. EXAMPLE: "We found fifteen 2-ft pieces of pipe in the machinery."

Rule 8: Only Rarely Use Numbers in Parentheses

Except in legal documents, avoid the practice of placing figures in parentheses after their word equivalents. EXAMPLE: "The second party will send the first party forty-five (45) barrels on or before the first of each month." Note that the parenthetical amount is placed immediately after the figure, not after the unit of measurement.

Rule 9: Use Figures with Dollars

Use figures with all dollar amounts, with the exception of the context noted in Rule 8. Avoid cents columns unless exactness to the penny is necessary.

Rule 10: Use Commas in Four-Digit Figures

To prevent possible misreading, use commas in figures of four digits or more. EXAMPLES: 15,000; 1,247; 6,003.

Rule 11: Use Words for Ordinals

Usually spell out the ordinal form of numbers. EXAMPLE: "The government informed all parties of the *first, second,* and *third* [not *1st, 2nd,* and *3rd*] choices in the design competition." A notable exception is tables and figures, where space limitations could argue for the abbreviated form.

Oral/Verbal

Oral refers to words that are spoken, as in "oral presentation." The term *verbal* refers to spoken or written language. To prevent confusion, avoid the word *verbal* and instead specify your meaning with the words *oral* and *written.* EXAMPLES:

- In its international operations, Digital Essentials has learned that some countries still rely on *oral* [not *verbal*] contracts.
- Their *oral* agreement last month was followed by a *written* [not *verbal*] contract this month.

Parts of Speech

The term *parts of speech* refers to the eight main groups of words in English grammar. A word's placement in one of these groups is based on its function within the sentence.

Noun. Words in this group name persons, places, objects, or ideas. The two major categories are (1) proper nouns and (2) common nouns. Proper nouns name specific persons, places, objects, or ideas, and they are capitalized. EXAMPLES: Cleveland; Mississippi River; Service Solutions, Inc.; Student Government Association; Susan Jones; Existentialism. Common nouns name general groups of persons, places, objects, and ideas, and they are not capitalized. EXAMPLES: trucks, farmers, engineers, assembly lines, philosophy.

Verb. A verb expresses action or state of being. Verbs give movement to sentences and form the core of meaning in your writing. EXAMPLES: explore, grasp, write, develop, is, has.

Pronoun. A pronoun is a substitute for a noun. Some sample pronoun categories include (1) personal pronouns (I, we, you, she, he), (2) relative pronouns (who, whom, that, which), (3) reflexive and intensive pronouns (myself, yourself, itself), (4) demonstrative pronouns (this, that, these, those), and (5) indefinite pronouns (all, any, each, anyone).

Adjective. An adjective modifies a noun. EXAMPLES: horizontal, stationary, green, large, simple.

Adverb. An adverb modifies a verb, an adjective, another adverb, or a whole statement. EXAMPLES: soon, generally, well, very, too, greatly.

Preposition. A preposition shows the relationship between a noun or pronoun (the object of a preposition) and another element of the sentence. Forming a prepositional phrase, the preposition and its object can reveal relationships, such as location ("They went *over the hill*"), time ("He left *after the meeting*"), and direction ("She walked *toward the office*").

Conjunction. A conjunction is a connecting word that links words, phrases, or clauses. EXAMPLES: and, but, for, nor, although, after, because, since.

Interjection. As an expression of emotion, an interjection can stand alone ("Look out!") or can be inserted into another sentence.

Passed/Past

Passed is the past tense of the verb *pass*, whereas *past* is an adjective, a preposition, or a noun that means "previous" or "beyond" or "a time before the present." EXAMPLES:

- He *passed* the survey marker on his way to the construction site.
- The *past* president attended last night's meeting. [adjective]

- He worked *past* midnight on the project. [preposition]
- In the distant *past*, the valley was a tribal hunting ground. [noun]

Per

Coming from the Latin, *per* should be reserved for business and technical expressions that involve statistics or measurement—such as *per annum* or *per mile*. It should not be used as a stuffy substitute for "in accordance with." EXAMPLES:

- Her *per diem* travel allowance of $90 covered hotels and motels.
- During the oil crisis years ago, gasoline prices increased by more than 50 cents *per gallon*.
- *As you requested* [not *per your request*], we have enclosed brochures on our products.

Per Cent/Percent/Percentage

Per cent and *percent* have basically the same usage and are used with exact numbers. The one word *percent* is preferred. Even more common in technical writing, however, is the use of the percent sign (%) after numbers. The word *percentage* is only used to express general amounts, not exact numbers. EXAMPLES:

- After completing a marketing survey, Heavy Construction, Inc., discovered that 83 *percent* [or 83%] of its current clients have hired Heavy Construction for previous projects.
- A large *percentage* of the defects can be linked to the loss of two experienced quality-control inspectors.

Practical/Practicable

Although close in meaning, these two words have quite different implications. *Practical* refers to an action that is known to be effective. *Practicable* refers to an action that can be accomplished or put into practice, without regard for its effectiveness or practicality. EXAMPLES:

- His *practical solution* to the underemployment problem led to a 30% increase in employment last year.
- The department head presented a *practicable* response, because it already had been put into practice in another branch.

Principal/Principle

When these two words are misused, the careful reader notices. Keep them straight by remembering this simple distinction: *Principle* is always a noun that means "basic truth, belief, or theorem." EXAMPLE: "He believed in the principle of free speech." *Principal* can be either a noun or an adjective and has three basic uses:

- **As a noun meaning "head official" or "person who plays a major role."** EXAMPLE: We asked that a *principal* in the firm sign the contract.

- **As a noun meaning "the main portion of a financial account upon which interest is paid."** EXAMPLE: If we deposit $5,000 in *principal*, we will earn 9 percent interest.

- **As an adjective meaning "main or primary."** EXAMPLE: We believe that the *principal* reason for contamination at the site is the leaky underground storage tank.

Pronouns: Agreement and Reference

A pronoun is a word that replaces a noun, which is called the *antecedent* of the pronoun. EXAMPLES: this, it, he, she, they. Pronouns, as such, provide you with a useful strategy for varying your style by avoiding repetition of nouns. Following are some rules to prevent pronoun errors:

Rule 1: Make Pronouns Agree with Antecedents

Check every pronoun to make certain it agrees with its antecedent in number—that is, both noun and pronoun must be singular, or both must be plural. Of special concern are the pronouns *it* and *they*. EXAMPLES:

- Change "DigiCorps plans to complete their Argentina project next month" to this sentence: "DigiCorps plans to complete its Argentina project next month."

- Change "The committee released their recommendations to all departments" to this sentence: "The committee released its recommendations to all departments."

Rule 2: Be Clear About the Antecedent of Every Pronoun

There must be no question about what noun a pronoun replaces. Any confusion about the antecedent of a pronoun can change the entire meaning of a sentence. To avoid such reference problems, it may be necessary to rewrite a sentence or even use a noun rather than a pronoun. Do whatever is necessary to prevent misunderstanding by your reader. EXAMPLE: Change "The gas filters for these tanks are so dirty that they should not be used" to this sentence: "These filters are so dirty that they should not be used."

Rule 3: Avoid Using *This* as the Subject Unless a Noun Follows It

A common stylistic error is the vague use of *this*, especially as the subject of a sentence. Sometimes the reference is not clear at all; sometimes the reference may be clear after several readings. In almost all cases, however, the use of *this* as a pronoun reflects poor technical style and tends to make the reader want to ask, "This what?" Instead, make the subject of your sentences concrete, either by adding a noun after the *this* or by recasting the sentence. EXAMPLE: Change "He talked constantly about the project to be completed at the Olympics. This made his office-mates irritable" to the following: "His constant talk about the Olympics project irritated his office-mates."

Punctuation: General

Commas.   Most writers struggle with commas, so you are not alone. The problem is basically threefold. First, the teaching of punctuation has been approached in different, and sometimes quite contradictory, ways. Second, comma rules themselves are subject to various interpretations. And third, problems with comma placement often mask more fundamental problems with the structure of a sentence itself.

Start by knowing the basic rules of comma use. The rules that follow are fairly simple. If you learn them now, you will save yourself a good deal of time later because you will not be questioning usage constantly. In other words, the main benefit of learning the basics of comma use is increased confidence in your own ability to handle the mechanics of editing. (If you do not understand some of the grammatical terms that follow, such as *compound sentence,* refer to the section on sentence structure.)

Rule 1: Commas in a Series

Use commas to separate words, phrases, and short clauses written in a series of three or more items. EXAMPLE: "The samples contained gray sand, sandy clay, and silty sand." According to current U.S. usage, a comma always comes before the "and" in a series. (In the United Kingdom, the comma is left out.)

Rule 2: Commas in Compound Sentences

Use a comma before the conjunction that joins main clauses in a compound sentence. EXAMPLE: "We completed the drilling at the Smith Industries location, and then we grouted the holes with Sakrete." The comma is needed here because it separates two complete clauses, each with its own subject and verb (*we completed* and *we grouted*). If the second *we* had been deleted, there would be only one clause containing one subject and two verbs ("we completed and grouted"). Thus no comma would be needed. Of course, it may be that a sentence following this comma rule is far too long; do not use the rule to string together intolerably long sentences.

Rule 3: Commas with Nonessential Modifiers

Set off nonessential modifiers with commas at the beginning, middle, or end of sentences. *Nonessential modifiers* are usually phrases that add more information to a sentence, rather than greatly changing its meaning. When you speak, there is often a pause between this kind of modifier and the main part of the sentence, giving you a clue that a comma break is needed. EXAMPLE: "The report, which we submitted three weeks ago, indicated that the company would not be responsible for transporting hazardous wastes." But—"The report that we submitted three weeks ago indicated that the company would not be responsible for transporting hazardous wastes." The first example includes a nonessential modifier, would be spoken with pauses, and therefore uses separating commas. The second example includes an essential modifier, would be spoken without pauses, and therefore includes no separating commas.

Rule 4: Commas with Adjectives in a Series

Use a comma to separate two or more adjectives that modify the same noun at the same level of detail. To help you decide if adjectives modify the same noun equally, use this test: If you can reverse their positions and still retain the same meaning, then the adjectives modify the same word and should be separated by a comma. EXAMPLE: "Jason found the old, rotted gaskets."

Rule 5: Commas with Introductory Elements

Use a comma after introductory phrases or clauses of about five words or more. EXAMPLE: "After completing the topographic survey of the area, the crew returned to headquarters for its weekly project meeting." Commas like the one after *area* help readers separate secondary or modifying points from your main idea, which of course should be in the main clause. Without these commas, there may be difficulty reading such sentences properly.

Rule 6: Commas in Dates, Titles, Etc.

Abide by the conventions of comma usage in punctuating dates, titles, geographic place names, and addresses. EXAMPLES:

- May 3, 2006, is the projected date of completion. (However, note the change in the military form of dates: We will complete the project on 3 May 2006.)
- John F. Dunwoody, Ph.D., has been hired to assist on the project.
- Heavy Construction has been selected for the project.
- He listed Dayton, Ohio, as his permanent residence.

Note the need for commas after the year *2006*, the title *Ph.D.*, the designation Inc., and the state name *Ohio*. Also note that if the day had not been in the first example, there would be no comma between the month and year and no comma after the year.

Semicolons. The semicolon is easy to use if you remember that it, like a period, indicates the end of a complete thought. Its most frequent use is in situations where grammar rules would allow you to use a period but where your stylistic preference is for a less abrupt connector. EXAMPLE: "Five engineers left the convention hotel after dinner; only two returned by midnight."

One of the most common punctuation errors, the comma splice, occurs when a comma is used instead of a semicolon or period in compound sentences connected by words, such as *however, therefore, thus,* and *then*. When you see that these connectors separate two main clauses, make sure either to use a semicolon or to start a new sentence. EXAMPLE: "We made it to the project site by the agreed-on time; however, [or "…time. However,…"] the rain forced us to stay in our trucks for two hours."

As noted in the "Lists" entry, there is another instance in which you might use semicolons. Place them after the items in a list when you are treating the list like a sentence and when any one of the items contains internal commas.

Colons. As mentioned in the "Lists" entry, you should place a colon immediately after the last word in the lead-in before a formal list of bulleted or numbered items. EXAMPLE: "Our field study involved these three steps:" or "In our field study, we were asked to:" The colon may come after a complete clause, as in the first example, or it may split a grammatical construction, as in the second example. However, it is preferable to use a complete clause before a formal list.

The colon can also be used in sentences in which you want a formal break before a point of clarification or elaboration. EXAMPLE: "They were interested in just one result: quality construction." In addition, use the colon in sentences in which you want a formal break before a series that is not part of a listing. EXAMPLE: "They agreed to perform all on-site work required in these four cities: Houston, Austin, Laredo, and Abilene." However, note that there is no colon before a sentence series without a break in thought. EXAMPLE: "They agreed to perform all the on-site work required in Houston, Austin, Laredo, and Abilene."

Apostrophes.   The apostrophe can be used for contractions, for some plurals, and for *possessives*. Only the last two uses cause confusion. Use an apostrophe to indicate the plural form of a word as a word. EXAMPLE: "That redundant paragraph contained seven *area*'s and three *factor*'s in only five sentences." Although some writers also use apostrophes to form the plurals of numbers and full-cap abbreviations, the current tendency is to include only the *s*. EXAMPLES: 7s, ABCs, PCBs, P.E.s.

As for possessives, you probably already know that the grammar rules seem to vary, depending on the reference book you are reading. Following are some simple guidelines:

Possessive Rule 1

Form the possessive of multisyllabic nouns that end in *s* by adding just an apostrophe, whether the nouns are singular or plural. EXAMPLES: actress' costume, genius' test score, the three technicians' samples, Jesus' parables, the companies' joint project.

Possessive Rule 2

Form the possessive of one-syllable, singular nouns ending in *s* or an *s* sound by adding an apostrophe plus *s*. EXAMPLES: Hoss's horse, Tex's song, the boss's progress report.

Possessive Rule 3

Form the possessive of all plural nouns ending in *s* or an *s* sound by adding just an apostrophe. EXAMPLES: the cars' engines, the ducks' flight path, the trees' roots.

Possessive Rule 4

Form the possessive of all singular and plural nouns not ending in *s* by adding an apostrophe plus *s*. EXAMPLES: the man's hat, the men's team, the company's policy.

Possessive Rule 5

Form the possessive of paired nouns by first determining whether there is joint ownership or individual ownership. For joint ownership, make only the last noun possessive. For individual ownership, make both nouns possessive. EXAMPLE: "Susan and Terry's project was entered in the science fair; but Tom's and Scott's projects were not."

Quotation Marks.   In technical writing, you may want to use this form of punctuation to draw attention to particular words, to indicate passages taken directly from another source, or to enclose the titles of short documents such as reports or book chapters. The rule to remember is this: Periods and commas go inside quotation marks; exclamation marks, question marks, semicolons, and colons go outside quotation marks.

Parentheses.   Use parentheses carefully, because long parenthetical expressions can cause the reader to lose the train of thought. This form of punctuation can be used when you (1) place an abbreviation after a complete term, (2) add a brief explanation within the text, or (3) include reference citations within the document text (as explained in Chapter 14). The period goes after the closing parenthesis when the parenthetical information is part of the sentence, as in the previous sentence. (However, it goes inside the closing parenthesis when the parenthetical information forms its own sentence, as in the sentence you are reading.)

Brackets.   Use a pair of brackets for the following purposes: (1) to set off parenthetical material already contained within another parenthetical statement and (2) to draw attention to a comment you are making within a quoted passage. EXAMPLE: "Two Heavy Construction studies have shown that the Colony Dam is up to safety standards. (See Figure 4-3 [Dam Safety Record] for a complete record of our findings.) In addition, the county engineer has a letter on file that will give further assurance to prospective home-owners on the lake. His letter notes that 'After finishing my three-month study [he completed the study in July 2007], I conclude that the Colony Dam meets all safety standards set by the county and state governments.'"

Hyphens.   The hyphen is used to form certain word compounds in English. Although the rules for its use sometimes seem to change from handbook to handbook, those that follow are the most common:

Hyphen Rule 1
Use hyphens with compound numerals. EXAMPLE: twenty-one through ninety-nine.

Hyphen Rule 2
Use hyphens with most compounds that begin with *self.* EXAMPLES: self-defense, self-image, self-pity. Other *self* compounds, like *selfhood* and *selfsame,* are written as unhyphenated words.

Hyphen Rule 3
Use hyphens with group modifiers when they precede the noun but not when they follow the noun. EXAMPLES: A well-organized paper, a paper that was well organized, twentieth-century geotechnical technology, bluish-gray shale, fire-tested material, thin-bedded limestone.

However, remember that when the first word of the modifier is an adverb ending in *—ly,* place no hyphen between the words. EXAMPLES: carefully drawn plate, frightfully ignorant teacher.

Hyphen Rule 4

Place hyphens between prefixes and root words in the following cases: (1) between a prefix and a proper name (ex-Republican, pre-Sputnik); (2) between some prefixes that end with a vowel and root words beginning with a vowel, particularly if the use of a hyphen would prevent an odd spelling (semi-independent, re-enter, re-elect); and (3) between a prefix and a root when the hyphen helps to prevent confusion (re-sent, not resent; re-form, not reform; re-cover, not recover).

Punctuation: Lists

As noted in Chapter 4 ("Page Design"), listings draw attention to parallel pieces of information whose importance would be harder to grasp in paragraph format. In other words, use lists as an attention-getting strategy. Following are some general pointers for punctuating lists. (See pages 46–48 in Chapter 3 for other rules for lists.)

You have three main options for punctuating a listing. The common denominators for all three are that you (1) always place a colon after the last word of the lead-in and (2) always capitalize the first letter of the first word of each listed item.

Option A: Place no punctuation after listed items.   This style is appropriate when the list includes only short phrases. More and more writers are choosing this option, as opposed to option B. EXAMPLE:

In this study, we will develop recommendations that address these six concerns in your project:

- Site preparation
- Foundation design
- Sanitary-sewer design
- Storm-sewer design
- Geologic surface faulting
- Projections for regional land subsidence.

Option B: Treat the list like a sentence series.   In this case, you place commas or semicolons between items and a period at the end of the series. Whether you choose option A or B largely depends on your own style or that of your employer. EXAMPLE:

In this study, we developed recommendations that dealt with four topics:

- Site preparation,
- Foundation design,
- Sewer construction, and
- Geologic faulting.

Note that this option requires you to place an *and* after the comma that appears before the last item. Another variation of option B occurs when you have internal commas within one or more of the items. In this case, you must change the commas that follow the listed items into semicolons. Yet you still keep the *and* before the last item. EXAMPLE:

Last month we completed environmental assessments at three locations:

- A gas refinery in Dallas, Texas;
- The site of a former chemical plant in Little Rock, Arkansas; and
- A waste pit outside of Baton Rouge, Louisiana.

Option C: Treat each item like a separate sentence.   When items in a list are complete sentences, you may want to punctuate each one like a separate sentence, placing a period at the end of each. You *must* choose this option when one or more of your listed items contain more than one sentence. EXAMPLE:

The main conclusions of our preliminary assessment are summarized here:

- At five of the six borehole locations, petroleum hydrocarbons were detected at concentrations greater than a background concentration of 10 mg/kg.
- No PCB concentrations were detected in the subsurface soils we analyzed. We will continue the testing, as discussed in our proposal.
- Sampling and testing should be restarted three weeks from the date of this report.

Regrettably/Regretfully

Regrettably means "unfortunately," whereas *regretfully* means "with regret." When you are unsure of which word to use, substitute the definitions to determine correct usage. EXAMPLES:

- *Regrettably*, the team members omitted their resumes from the proposal.
- Hank submitted his resume to the investment firm, but, *regrettably*, he forgot to include a cover letter.
- I *regretfully* climbed on the plane to return home from Hawaii.

Respectively

Some good writers may use *respectively* to connect sets of related information. Yet such usage creates extra work for readers by making them reread previous passages. It is best to avoid *respectively* by rewriting the sentence, as shown in the several following options. EXAMPLES:

Original: Appendices A, G, H, and R contain the topographical maps for Sites 6, 7, 8, and 10, respectively.

Revision—Option 1:   Appendix A contains the topographical map for Site 6; Appendix G contains the map for Site 7; Appendix H contains the map for Site 8; and Appendix R contains the map for Site 10.

Revision—Option 2:   Appendix A contains the topographical map for Site 6; Appendix G for Site 7; Appendix H for Site 8; and Appendix R for Site 10.

Revision—Option 3: Topographical maps are contained in the appendices, as shown in the following list:

Appendix	Site
A	6
G	7
H	8
R	10

Set/Sit

Like *lie* and *lay*, *sit* and *set* are verbs distinguished by form and use. Following are the basic differences:

1. *Set* means "to place in a particular spot" or "to adjust." It is a transitive verb and thus takes a direct object to which it conveys action. Its main parts are *set* (present), *set* (past tense), *set* (past participle), and *setting* (present participle).

2. *Sit* means "to be seated." It is usually an intransitive verb and thus does not take a direct object. Its main parts are *sit* (present), *sat* (past), *sat* (past participle), and *sitting* (present participle). It can be transitive when used casually as a direction to be seated. ("Sit yourself down and take a break.")

Examples:

■ He *set* the computer on the table yesterday.

■ While *setting* the computer on the table, he sprained his back.

■ The technician had *set* the thermostat at 75 degrees.

■ She plans to *sit* exactly where she sat last year.

■ While *sitting* at her desk, she saw the computer.

Sic

Latin for "thus," *sic* is most often used when a quoted passage contains an error or other point that might be questioned by the reader. Inserted within brackets, *sic* shows the reader that the error was included in the original passage and that it was not introduced by you. EXAMPLE: "The customer's letter to our sales department claimed that 'there are too [*sic*] or three main flaws in the product.'"

Spelling

All writers find at least some words difficult to spell, and some writers have major problems with spelling. Automatic spell-checking software helps solve the problem, but you must still remain vigilant during the proofreading stage. One or more misspelled words in an otherwise well-written document may cause readers to question professionalism in other areas.

However, you should keep your own list of words you most frequently have trouble spelling. Like most writers, you probably have a relatively short list of words that give you repeated difficulty.

Stationary/Stationery

Stationary means "fixed" or "unchanging," whereas *stationery* refers to paper and envelopes used in writing or typing letters. EXAMPLES:

- To perform the test correctly, one of the workers had to remain *stationary* while the other one moved around the job site.
- When she began her own business, Julie purchased *stationery* with her new logo on each envelope and piece of paper.

Subject–Verb Agreement

Subject–verb agreement errors are quite common in technical writing. They occur when writers fail to make the subject of a clause agree in number with the verb. EXAMPLE: "The nature of the diverse geologic deposits are explained in the report." (The verb should be *is,* because the singular subject is *nature.*)

Writers who tend to make these errors should devote special attention to them. Specifically, isolate the subjects and verbs of all the clauses in a document and make certain that they agree. Following are seven specific rules for making subjects agree with verbs:

Rule 1: Subjects Connected by *And* Take Plural Verbs

This rule applies to two or more words or phrases that, together, form one subject phrase. EXAMPLE: "The site preparation section and the foundation design portion of the report are to be written by the same person."

Rule 2: Verbs After *Either/Or* or *Neither/Nor* Agree with the Nearest Subject

Subject words connected by *either* and *or* (or *neither* and *nor*) confuse many writers, but the rule is very clear. Your verb choice depends on the subject nearest the verb. EXAMPLE: "He told his group that neither the three reports nor the proposal was to be sent to the client that week."

Rule 3: Verbs Agree with the Subject, Not with the Subjective Complement

Sometimes called a *predicate noun* or *adjective,* a subjective complement renames the subject and occurs after verbs such as *is, was, are,* and *were.* EXAMPLE: "The theme of our proposal is our successful projects in that region of the state." However, the same rule would permit this usage: "Successful projects in that part of the state are the theme we intend to emphasize in the proposal."

Rule 4: Prepositional Phrases Do Not Affect Matters of Agreement

As long as, in addition to, as well as, and *along with* are prepositions, not conjunctions. A verb agrees with its subject, not with the object of a prepositional phrase. EXAMPLE: "The manager of human resources, along with the personnel director, is supposed to meet with the three applicants."

Rule 5: Collective Nouns Usually Take Singular Verbs

Collective nouns have singular form but usually refer to a group of persons or things (e.g., *team, committee, crew*). When a collective noun refers to a group as a whole, use a singular verb. EXAMPLE: "The project crew was ready to complete the assignment." Occasionally, a collective noun refers to the members of the group acting in their separate capacities. In this case, either use a plural verb or, to avoid awkwardness, reword the sentence. EXAMPLE: "The crew were not in agreement about the site locations" or, "Members of the crew were not in agreement about the site locations."

Rule 6: Foreign Plurals Usually Take Plural Verbs

Although usage is gradually changing, most careful writers still use plural verbs with *data, strata, phenomena, media,* and other irregular plurals. EXAMPLE: "The data he asked for in the request for proposal are incorporated into the three tables."

Rule 7: Indefinite Pronouns Like *Each* and *Anyone* Take Singular Verbs

Writers often fail to follow this rule when they make the verb agree with the object of a prepositional phrase instead of with the subject. EXAMPLE: "Each of the committee members are ready to adjourn" (incorrect). "Each of the committee members is ready to adjourn" (correct).

To/Too/Two

To is part of the infinitive verb form or is a preposition in a prepositional phrase. *Too* is an adverb that suggests an excessive amount or that means "also." *Two* is a noun or an adjective that stands for the numeral "2." EXAMPLES:

- He volunteered *to* go [infinitive verb] *to* Alaska [prepositional phrase] *to* work [another infinitive verb form] on the project.
- Stephanie explained that the proposed hazardous-waste dump would pose *too* many risks *to* the water supply. Scott made this point, *too*.

Utilize/Use

Utilize is simply a long form for the preferred verb "use." Although some verbs that end in —*ize* are useful words, most are simply wordy substitutes for shorter forms. As some writing teachers say, "Why use 'utilize' when you can use 'use'?"

Which/That

Which is used to introduce nonrestrictive clauses, which are defined as clauses not essential to meaning (as in this sentence). Note that such clauses require a comma before the *which* and a slight pause in speech. *That* is used to introduce restrictive clauses that are essential to the meaning of the sentence (as in this sentence). Note that such clauses have no comma before the *that* and are read without a pause. *Which* and *that* can produce different meanings, as in the following examples:

- Our benefits package, *which* is the best in our industry, includes several options for medical care.
- The benefits package *that* our firm provides includes several options for medical care.
- My daughter's school, *which* is in Cobb County, has an excellent math program.
- The school *that* my daughter attends is in Cobb County and has an excellent math program.

Note that the preceding examples with *that* might be considered wordy by some readers. Indeed, such sentences often can be made more concise by deleting the *that* introducing the restrictive clause. However, delete *that* only if you can do so without creating an awkward and choppy sentence.

Who/Whom

Who and *whom* give writers (and speakers) fits, but the importance of their correct use probably has been exaggerated. If you want to be one who uses them properly, remember this basic point: *Who* is a subjective form that can only be used in the subject slot of a clause; *whom* is an objective form that can only be used as a direct object or other nonsubject noun form of a sentence. You can check which word you should use by substituting *he* and *him*. Use *who* when you would use *he* and use *whom* when you would use *him*. EXAMPLES:

- The man *who* you said called me yesterday is a good customer of the firm. (The clause "who…called me yesterday" modifies "man." Within this clause, *who* is the subject of the verb "called." Note that the subject role of *who* is not affected by the two words "you said," which interrupt the clause.)

- They could not remember the name of the person *whom* they interviewed. (The clause "whom they interviewed" modifies "person." Within this clause, *whom* is the direct object of the verb "interviewed.")

Who's/Whose

Who's is a contraction that replaces *who is*, whereas *whose* is a possessive adjective. EXAMPLES:

- *Who's* planning to attend the annual meeting?
- Susan is the manager *who's* responsible for training.
- *Whose* budget includes training?
- Susan is the manager *whose* budget includes training.

Your/You're

Your is an adjective that shows ownership, whereas *you're* is a contraction for *you are*. EXAMPLES:

- *Your* office will be remodeled next week.
- *You're* responsible for giving performance appraisals.

Appendix B mytechcommlab

>>> Resources in MyTechCommLab
www.mytechcomlab.com

MyTechCommLab is an online resource of tutorials, model documents, and activities. Some of these materials are referred to in marginal notes throughout this textbook; this appendix provides a more complete list of materials available online.

Tutorials

Writing process

Writing a formal report

Writing and visuals

Visual design

Web design

Model Documents

Letters

Memos

E-mails

Career correspondence

Proposals

Abstracts

Reports

Instructions

Procedures

Descriptions

Definitions

Web sites

Presentations

Brochures

Technical marketing materials

Activities

Grammar diagnostics and exercises

Writing letters

Writing memos

Writing e-mails

Writing career correspondence

Writing proposals

Writing abstracts

Creating brochures

Writing short/informal reports

Writing formal reports

Writing instructions

Writing procedures

Writing definitions

Writing descriptions

Patterns of organization

Usability case studies

Other Resources

The Research Process

- MySearchLab™
- The Research Assignment
- Avoiding Plagiarism
- Student Bookshelf
- Research Writing Samples

Longman Online Handbook

Web Links to resources for technical communicators

Student Bookshelf of reference materials

Photo Credits

Chapter 1

Ryan McVay/Getty Images—Photodisc—Royalty Free, pp. 1, 13; Zefa Collection/CORBIS—NY, p. 20.

Chapter 2

Getty Images—Stockbyte, p. 28.

Chapter 3

© Bettmann/CORBIS All Rights Reserved, p. 43; Spencer Grant/PhotoEdit Inc., p. 45; Keith Brofsky/Getty Images Inc.—Stone Allstock, p. 46; © Dorling Kindersley, p. 53.

Chapter 4

Getty Images—Stockbyte, Royalty Free, p. 75; Photos.com, p. 76; Steve Gorton (c) Dorling Kindersley, p. 81; Hiep Vu/Masterfile Stock Image Library, p. 86.

Chapter 5

Nick Koudis/Getty Images, Inc.—Photodisc./Royalty Free, p. 93; Photodisc/Getty Images, p. 95; Superstock Royalty Free, p. 103.

Chapter 6

Photos.com, p. 108; Andrew Olney/Getty Images/Digital Vision, p. 109; Keith Brofsky/Getty Images, Inc.— Photodisc./Royalty Free, p. 115.

Chapter 7

EyeWire Collection/Getty Images—Photodisc-Royalty Free, pp. 123, 154; Photolibrary.com, p. 125; Medford Taylor/National Geographic Image Collection, p. 137; John A. Rizzo/Getty Images, Inc.—Photodisc./Royalty Free, p. 157; Tony Camacho/Photo Researchers, Inc. p. 159.

Chapter 8

Getty Images—Stockbyte, Royalty Free, p. 160; EyeWire Collection/Getty Images—Photodisc-Royalty Free, p. 163.

Chapter 9

Getty Images—Digital Vision, p. 189; Britt Erlanson/Image Bank/Getty Images, p. 194; Michael Matisse/Getty Images, Inc.—Photodisc, p. 195; Craig Brewer/Getty Images, Inc.—Photodisc, p. 196.

Chapter 10

Stockbyte/Getty Images—Stockbyte, Royalty Free, p. 201; James Woodson/Getty Images/Digital Vision, p. 202; Getty Images, Inc—Stockbyte Royalty Free, p. 215.

Chapter 11

Getty Images—Stockbyte, p. 222.

Index